Life Beyond Human Boundaries: John Dupré's Philosophy and the Interdisciplinary Study of the Life Sciences

Sabina Leonelli · Celso Neto · Stephan Guttinger
Editors

Life Beyond Human Boundaries: John Dupré's Philosophy and the Interdisciplinary Study of the Life Sciences

Springer

Editors
Sabina Leonelli
School of Social Sciences and Technology
Technical University of Munich
München, Bayern, Germany

Celso Neto
Centre for the Study of Life Sciences
University of Exeter
Exeter, UK

Stephan Guttinger
Centre for the Study of Life Sciences
University of Exeter
Exeter, UK

ISBN 978-3-032-26976-8 ISBN 978-3-032-26977-5 (eBook)
https://doi.org/10.1007/978-3-032-26977-5

This work was supported by University of Exeter and Technical University of Munich.

This Springer imprint is published by the registered company Springer Nature Switzerland AG
The registered company address is: Gewerbestrasse 11, 6330 Cham, Switzerland

Synopsis

John Dupré is internationally regarded as one of the most prominent living philosophers of science. His work has had a profound impact on philosophy and biology over the last half century. From fundamental questions about classification to metaphysical debates about the nature of organisms and life, Dupre's writings have changed how scholars look at the life sciences and the natural world. But it is not just his published output that has re-shaped the field. It is also his philosophical methodology, characterised by a naturalistic and interdisciplinary approach. In part this approach was shaped by Dupré's interactions with natural and social scientists, an aspect of his approach to philosophy that is also reflected in his role as a co-founder and then Director of the Egenis Centre for the Study of Life Sciences. Dupré's methodological approach has inspired philosophers to work across disciplinary boundaries and reflect carefully upon the question of how interdisciplinary work should be conducted.

This book provides an authoritative assessment of Dupré's various contributions to philosophy. It does so by bringing together leading philosophers, historians, and scientists to explore different aspects of Dupré's ideas in light of their own cutting-edge work. In this way, the book advances the state-of-the-art in philosophy, by scrutinising Dupré's highly influential views alongside those of other leading scholars. The result is a unique and often surprising multi-disciplinary perspective on Dupré's pivotal role in shaping the intellectual life and research methods of contemporary philosophy and future scholarship.

The book is divided into three parts, each reflecting an important theme in Dupré's work.

Part I considers the theme of promiscuity and disorder in nature. One of Dupré's most important contributions is the realisation that nature is complex and disunified in ways that require us to rethink both the nature of science and the nature of human beings. The section starts with consideration of the relation between classification and disunity, with Catherine Kendig discussing classificatory practices in contemporary agriculture, arguing that such practices carry a deep metaphysical commitment to process ontology. Nancy Cartwright's chapter examines the view that economics is not an exact science and only capable of providing limited causal knowledge.

Adrian Haddock examines the place of mind in this complex nature, exploring a tension between Dupré's empiricism and anti-reductionism.

Part II focuses on process ontology. For more than a decade, Dupré has been a staunch defender of processualism, reintroducing this ancient position in the philosophy of biology. Grounding his processualism in scientific theory and practice, Dupré has developed a framework for understanding life, individuality, and other natural phenomena. Ford Doolittle discusses the processual nature of evolution, exploring its implications for the theory of natural selection. Scott Gilbert shows how Dupré's process ontology fits with our current scientific understanding of symbiosis, metabolism, and microbial interactions. The topic of symbiosis and microbial complexity is also the centre of Thomas Pradeu's contribution. Pradeau argues that philosophers should work closely with scientists to understand that complexity, an approach exemplified by Dupré's work. Finally, Tim Lewens examines the impact of process ontology on the idea of human nature. Lewens defends a sceptical view of human nature by showing how processual our species is.

Part III reflects on the role of interdisciplinarity in philosophy and science. As the other themes of the volume show, John Dupré's work has been characterised by a determination to work across disciplinary boundaries. Jane Calvert, Robert Smith, and Erika Szymanski consider when (and how) interdisciplinarity can generate productive and successful work. Aligned with John's processualism, the authors reflect on what it would mean to collaborate with the very organisms that science studies. Jim Griesemer's chapter focuses on the collaboration among scientists and philosophers, considering how the tensions and frictions between these groups are productive in themselves. Alan Love reflects on the challenges and opportunities faced by philosophers and biologists collaborating on a major interdisciplinary project. Love focuses on a project around the topic of teleology. Finally, Phillip Kitcher turns to a key tension in Dupré's work, between promiscuous pluralism and ontological monism, and offers both a (pragmatist) way out from this tension and a heartfelt homage to Dupre's intellectual trajectory over the course of his career and their personal friendship.

Contents

About the Editors

Sabina Leonelli is professor of philosophy and history of science and technology at the Technical University of Munich, where she co-directs the Ethical Data Initiative and the Public Science Lab, and leads a large project on the philosophy of open science (http://www.opensciencestudies.eu). Until 2024 she directed the Exeter Centre for the Study of Life Sciences at the University of Exeter, where she holds a honorary professorship. Her research investigates the roles played by technology, data, models and collaborative methods within the life sciences and public health. She aims to contribute to an understanding of knowledge production that supports equitable, responsible, engaged and sustainable research practices. She is President of the International Society for the History, Philosophy and Social Studies of Biology; a Fellow of the American Association for the Advancement of Science, International Society for the Philosophy of Science, Academia Europaea and the Royal Society of Biology; and winner of the Lakatos Award (2018) and Patrick Suppes Prize in Philosophy of Science (2022).

Celso Neto is associate professor in philosophy at the University of Exeter, where he is a member of the Centre for the Study of Life Sciences and leads a project on race and genomics funded by the European Research Council. Before joining Exeter, Celso held a postdoctoral fellowship at Dalhousie University, working across the philosophy and biochemistry and molecular biology departments. His main interest lies in understanding how biases and values influence methodologies and reasoning in science, with a particular focus on evolutionary biology and its intersection with societal issues.

Stephan Guttinger is a lecturer in philosophy at the University of Exeter, UK. He is co-lead of the *Data, Knowledge and AI* research strand at the Egenis Centre for the Study of Life Sciences and also serves as the head of research for the Exeter strand of the *Ethical Data Initiative,* a project jointly hosted by the University of

Exeter and the Technische Universität München (TUM). Stephan's work focuses on knowledge production in the biological laboratory, with a particular focus on issues such as experimental control, reproducibility, and the increasing use of AI-driven automation.

Chapter 1
Introduction: Delving into Dupré's Philosophy

Celso Neto, Stephan Guttinger, and Sabina Leonelli

Abstract This introduction presents an overview of central themes and positions developed by John Dupré since the 1980s. We focus on key innovations in his work, including pluralism in biological classification; processual thinking; and interdisciplinary methodology. By exploring these topics and their historical development in Dupré's work, we offer the reader a background for the remaining chapters in this volume and an overview of why such research continues to be foundational for the philosophy of biology and the philosophy of science more generally.

Over the course of several decades, John Dupré's philosophical work has played a key role in re-orienting philosophy of science, metaphysics, and philosophy more generally. His work generates philosophical insights from a deep engagement with cutting-edge science, drawing attention to the plurality of objects and practices in the sciences and their historical and social context. In proposing an innovative take on the long-standing problem of classification in his seminal monograph *The Disorder of Things*, Dupré provided a comprehensive critique of reductionism, determinism, and essentialism. This critique still shapes philosophy of biology, philosophy of science, and metaphysics to this day. His pleas for "promiscuous realism", ontological pluralism, and—more recently—process philosophy challenge entrenched views about the unity of science and the natural world, making it possible to understand reality as multiple, diversified, and ever-changing process, and to emphasise

C. Neto (✉) · S. Guttinger · S. Leonelli
Egenis Centre for the Study of Life Sciences, University of Exeter, Exeter, UK
e-mail: c.alves-neto@exeter.ac.uk

S. Guttinger
e-mail: s.m.guttinger@exeter.ac.uk

S. Leonelli
e-mail: sabina.leonelli@tum.de

S. Leonelli
Chair of Philosophy and History of Science and Technology, Department of Science, Technology and Society, Technical University of Munich, Munich, Germany

S. Leonelli et al. (eds.), *Life Beyond Human Boundaries: John Dupré's Philosophy and the Interdisciplinary Study of the Life Sciences*,
https://doi.org/10.1007/978-3-032-26977-5_1

how methods of inquiry may need to suit such dynamic features. In this way, Dupré played a key role in showing how biology may constitute a worthwhile and significant source of philosophical insight about science as a whole—a ground-breaking achievement, especially given the role traditionally assigned to theoretical physics as the main model for scientific reasoning.

This volume offers a systematic and critical analysis of Dupré's work. It does so by bringing together leading philosophers, historians, and scientists who engage with different aspects of Dupré's ideas and assess them in relation to their own cutting-edge work. In this way, the volume advances the state-of-the-art in philosophy, scrutinising Dupré's highly influential views and putting them in dialogue with other leading scholars, insights, and findings. The result is a unique and multidisciplinary perspective on central topics that have dominated philosophy of science for decades. Thus, the volume also captures the evolution of these topics, as well as the methodological re-orientation of philosophy inspired by Dupré's work.

Each chapter in this volume critically engages with one or more key philosophical ideas proposed by Dupré. To ease the reader into these chapters, this introduction offers an overview of Dupré's contributions to general philosophy and philosophy of science, as well as an assessment of how his views have evolved over time. Not only do we showcase the enormous diversity of philosophical interests and the variety of biological domains covered by his work, but we also emphasize Dupré's deep engagement with a wide variety of biological domains that underpin his work. To give a flavor of this diversity: Dupré has contributed new perspectives to biological debates about species; human nature; individuality; the microbial world; the boundaries between living systems and non-living objects; and sex and gender (among other topics). Few philosophers have delved into so many characteristics of life and living organisms, while at the same time engaging rigorously with cutting-edge developments in areas as diverse as evolutionary theory, taxonomy, developmental biology, systems biology, synthetic biology, microbiology, plant science, biomedicine, and—of course—genomics. While this introduction cannot cover all these topics and areas, it will walk the reader through some of the key milestones.

1.1 Disorder of Nature, Disorder of Science

Dupré's early work revolves around three main interrelated topics: classificatory pluralism, anti-essentialism, and anti-reductionism. This section presents these topics, offering an overview of how Dupré explores them to show the complexity of science and nature. This overview provides a background to the chapters in PART I of this volume, which challenge that understanding.

We start with issues around classification. Dupré's interest in classificatory practices predates his philosophical work and persists throughout his career (Dupré, 2025a). He finds in biological classifications a rich source of philosophical analysis, a springboard to intervening in core debates in the philosophy of language, metaphysics, and the general philosophy of science.

In the early 1980s, the works of Saul Kripke (1980) and Hilary Putnam (1975) had become hugely influential in Anglo-American philosophy. This work renewed philosophical interest in natural kinds, i.e., the idea that nature is divided into groupings that reflect the structure of the natural world (Bird & Tobin, 2024). Each natural kind is defined by a real essence, i.e., a set of individually necessary and jointly sufficient properties distinctive of the kind. For example, water can be understood as a natural kind defined by the chemical composition H_2O. According to Putnam, natural kind terms (e.g., "water") refer to all and only the entities that have the real essences of that kind, regardless of whether speakers know about these essences. This view implies a strong continuity between scientific and ordinary language. In ordinary contexts, people without any knowledge of chemistry still use the term "water" to pick out the very same entities identified by chemists as water.

Dupré's first publication challenged the continuity between scientific and ordinary language (Dupré, 1981). Attention to biological classification provides numerous examples of the conflict between scientific and non-scientific contexts. For example, when lay people refer to lilies, they only refer to some species and genera within the family *Liliaceae* (ibid., 74). This way of grouping organisms conflicts with biological taxonomy and systematics, according to which lilies would refer to the entire family *Liliaceae*. The hierarchical structure of biological taxonomy also creates challenges for the continuity between scientific and ordinary language. According to Putnam, the reference of a natural kind term is first determined by identifying a suitable exemplar of that kind, i.e., an organism that is a member of the kind. The other members of the kind are then determined by surveying all the other organisms that share the same 'sameness relation' (or real essence) with the first exemplar. The problem is that, for any exemplar, numerous sameness relations exist with other organisms across hierarchical levels. For example, the term "beetle" ordinarily applies to organisms from different species. If scientists try to determine the extension of that term, however, they will necessarily pick out an exemplar from one species or another. Putnam's theory would have to explain why scientists associate the term "beetle" with sameness relations between organisms from different species rather than between the exemplar and fellow members of the same species. Thus, science and the natural world seem too complex to assume the continuity between scientific and ordinary discourse.

The analysis of biological classification and ordinary kind terms paved the way for Dupré's scientifically informed metaphysical theory, namely *promiscuous realism* (Dupré, 1981, 1993a, 1996). This is one of Dupré's main contributions to philosophy. After recognizing that there are many 'sameness relations' among organisms, Dupré proposes that there is no single relation that can be considered superior or privileged when grouping organisms into the same species, genus, family, etc. Instead, there are many equally legitimate ways to generate classifications at all levels. This idea corresponds to the 'promiscuous' (or pluralist) element of Dupré's view. The 'realist' element of his view states that different sameness relations are biologically legitimate. These relations are based on the natural properties of organisms and serve to make biologically relevant distinctions between them, distinctions that matter for various human concerns. So understood, realism is not fully independent from human interests. Furthermore, Dupré does not necessarily privilege grouping practices in scientific taxonomy over non-scientific classifications. Famously, culinary classifications are expected to emphasize the distinction between garlic and onions, even though in scientific taxonomy these groups fall under the same genus (Dupré, 1993a). Meanwhile, taxonomists might be interested in distinguishing different species of frogs, but these distinctions are likely irrelevant in the culinary context.

Promiscuous realism is a radical pluralist view of classification with significant consequences in philosophy. Famously, biologist Michael Ghiselin (1974) and philosopher David Hull (1976) argued that species taxa are not examples of natural kinds but rather individuals, i.e., genealogical evolving lineages extended over space and time. Organisms are parts of a species in virtue of their genealogical connections rather than intrinsic properties (e.g., morphological or genetic properties). This position was becoming hegemonic in the 1990s, but Dupré (1993a) challenged one of its key premises. He rejected the idea that the nature of species is necessarily a function of evolutionary theory. If there is no privileged way of grouping organisms into the same species, and if biologists have many other research programs besides evolutionary theory, then the view that species are individuals is too restrictive. For example, ecologists working with predator-prey models might treat species as kinds rather than individuals. Species would be defined by ecological properties or roles exhibited by their members. Thus, promiscuous realism suggests that species taxa can be both individuals and natural kinds.

Promiscuous realism also intervenes in the question of how to define biological species. This question has vexed biologists and philosophers for decades (Ereshefsky, 1992, 2022). Numerous 'species concepts' have been proposed (Coyne & Orr, 2004). Some concepts treat species as reproductive isolated metapopulations, ecological communities, evolutionary lineages, or groups of organisms sharing an exclusive common ancestor (i.e., monophyletic groups). Despite all disagreements, many scholars tend to favour species concepts related to evolutionary principles or processes. Promiscuous pluralism has radically challenged this tendency. According to Dupré, there are too many biological processes carving up the natural world, and evolution-inspired species concepts cannot account for this variability (Dupré, 1993a, 1996). Here we can notice how Dupré's already hints at a view of the natural world as a complex set of processes, a topic that will occupy the center of his later career (Sects. 1.3 and 1.4). In his earlier work, however, Dupré discusses biological processes mainly as grounds for his promiscuous realism. In this volume, Catherine Kendig moves beyond this relationship, exploring how processes ground the normativity of scientific classifications (Chap. 2, Part I). Kendig investigates classifications of soil in US agricultural history to show how understanding soil as a process influence ethical and political uses of those classifications. In this way, Kendig invite us to consider how Dupré's views on metaphysics and classification can have normative consequences.

Dupré's views on biological classification and species have evolved in later publications (2001). By the late 1980s, the "cladistic" school of taxonomy had become dominant among systematists (Hull, 1988). This taxonomic approach understands that biological classification should capture the genealogical patterns of evolution, classifying organisms into specific evolutionary lineages. Again, Dupré challenged the status quo. According to him, the primary aim of classifications is to sort "organisms into kinds in ways that will serve to convey and store information about the biological world" (Dupré, 2001, 208). Classifications are supposed to be as comprehensive and stable as possible, but these desiderata frequently clash with the goals of cladistic taxonomy. In other words, the species that figure as basal units of classification might not coincide with the evolutionary lineages of interest to cladists and evolutionary-inclined systematists. Dupré concludes that one should not conflate these evolutionary lineages with those units of classification, proposing that one reserves the term "species" for the latter. This suggestion entails that, contrary to his previous work, Dupré would take "species" to apply only to natural kinds.

Traditionally, natural kinds are understood through the lens of kind essentialism (Okasha, 2002). Kind essentialism identifies a set of properties that determines the nature and extension of a natural kind. These properties must be instantiated by all and only members of that kind, and they must causally explain several other properties taken to be distinctive of the kind. Dupré is well-known for his vigorous rejection of this form of essentialism. He emphasizes the variability in nature and, thus, the trouble in distinguishing between the supposedly distinctive properties of a kind from mere accidental properties (Dupré, 1981, 1986, 1993a). For example, one might say that squirrels' genetic structure is the essence of being a squirrel, such that every organism with that genetic structure will also possess the distinctive phenotypic properties of the kind. However, it is expected that many squirrels will not have these distinctive properties (e.g., a fluffy tail). Given that all organismal properties are statistical, it is hard to identify in a non-question-begging way a set of distinctive properties associated with and explained by the supposedly underlying essence of the kind.

Dupré's anti-essentialism also puts into question intuitive notions of sex and gender (Dupré, 1986, 1993a, 2002). One obvious feature of the natural world is the distinction between males and females across different species. This distinction seems to involve differences in physiology, behaviour, and other levels of organization. According to kind essentialism, these differences are explained by a more fundamental difference in essential properties unique to each side of the divide. The problem is that it is hard to find a set of properties shared by all and only females (or males) that is causally explanatory. For instance, defining females as having two X chromosomes fails to apply across species. Another candidate definition is based on anisogamy, i.e., differences in gamete size between males and females. However, as Dupré points out, properties and causes related to the difference in gamete sizes can widely differ across species. Differences in chromosomal structure (XX versus XY) might cause anisogamy in humans, but these same chromosomal structures might be independently responsible for the physiological and behavioural differences between males and females. Thus, these differences are not explained by sex but rather by the cause of sex (Dupré, 1993a). The upshot is that sex categories such as 'male' and 'female' do not carry much explanatory power. A similar conclusion involves the gender categories of 'men' and 'women.' There is so much variability in gender behaviours across cultures that those categories cannot support wide-range generalizations about human behaviour.

Anti-essentialism is one of the motivations for another pillar of Dupré's philosophy, namely anti-reductionism (Dupré, 1993a). Different forms of reductionism have been popular in the philosophy of science and the philosophy of mind (Oppenheim & Putnam, 1958; Smart, 1959). In the philosophy of science, reductionism relates to the important discussion of the autonomy (and legitimacy) of non-physical sciences (Fodor, 1980). In this context, reductionism states that scientific knowledge is reducible to descriptions at the level of physics, i.e., as statements about particle arrangements and activities. Reductionism gives rise to a hierarchical view of the sciences, in which higher-level or 'special' sciences (chemistry, biology, psychology, etc.) can all be progressively recast in the language of physics. However, one reason to deny reductionism is that it presupposes essentialism. If reductionism implies the capacity to derive entities at one level (e.g., biology) from entities at a lower level (physics), there should be lawlike statements ("bridge principles") capable of connecting the entities at the two levels. These lawlike statements would have to enable us to identify entities belonging to a natural kind at the biological level solely based on information about the physical structure of those entities. In practice, this corresponds to spelling out the essential properties of that natural kind, an impossible task.

A motivation for reductionism is the idea of causal completeness in physics. According to this idea, every activity of physical entities is determined by the activities of their composing physical particles and the physical laws.[1] Thus, if an entity is fully made of physical particles, the behavior of these particles provides a full account of the entity's behaviour. The entity and its behaviour could then be reducible to its parts. While the idea of causal completeness was and still is widely shared by philosophers, there is simply no empirical evidence for it (Dupré, 2025a). Rather, Dupré raises the possibility that different levels and domains are partially autonomous and causally efficacious (Dupré, 1993a, 157). The natural kinds and causal explanations at one level do not necessarily map into kinds and causal explanations at a lower level. In this way, anti-reductionism is also vindicated.

Dupré's attack on reductionism goes beyond biology and physics (Dupré, 1993a). He also rejects the classical reductionism about the mind, according to which mental states are identical to and multiply realizable by lower-level (e.g., neurological) states. This reductionism suggests that ordinary linguistic expressions for mental states (e.g., "I love ice cream") are ultimately reducible to descriptions of those lower-level states. Dupré rejects this reductionism by highlighting the richness of human ordinary discourse about mental states (i.e., folk psychology). This discourse seems to contain all the resources needed to develop scientific descriptions, including neurobiological ones. In this volume, Adrian Haddock critically engages with this view, identifying a core tension within Dupré's thinking (Chap. 4, Part I). According to Haddock, if the ordinary language of mental states contains the resources for scientific language but is separate from it, that language cannot derive its meaning from empirical information. In other words, empiricism is at risk. Dupré is a self-declared empiricist in the sense that language and knowledge must be grounded in

[1] This idea can be described both in deterministic and indeterministic ways.

empirical content. However, if Haddock is correct, Dupré's anti-reductionism about the mental conflicts with that notion of empiricism.

Overall, Dupré's discussion of pluralism, anti-essentialism, and anti-reductionism offers a distinctive view of science and nature. Dupré challenges common sense views about nature (e.g., essentialism), which leads him to challenge common sense views about science (e.g., reductionism), and vice versa. Science and nature are pictured as fundamentally complex. This complexity is an invitation to rethink the limits of science, which Dupré considers one of the main tasks of philosophers of science. In his earlier work, Dupré takes on this task by examining the limits of economics (Dupré, 1993b). This science is traditionally portrayed as an 'exact' science, grounded in mathematical rigor and objective methodologies. However, the actual phenomena studied in economics are too complex as they are influenced by historical contingencies, variable human preferences, social institutions, etc. Economics cannot capture all this complexity, and only offers causal knowledge in a very loose sense. In Chap. 3 (PART I), Nancy Cartwright further develops this point, showing how the type of knowledge produced by economics is limited but useful. Cartwright argues that economics can offer valuable insights about the real world, but these insights are qualitative and inexact. Furthermore, these insights can only bear fruit if integrated with other sources of knowledge. This chapter nicely captures how many of the ideas of Dupré and Cartwright have evolved together since the early days of their careers when they were colleagues at Stanford University (Cartwright, 1983).

This picture of complex science opens the door to another important topic present in Dupré's work, namely the relationship between science and human values (Dupré, 1993a, 2002, 2007). If science is so complex and radically pluralistic, there are many legitimate ways to conduct scientific investigation. Knowledge will often be inexact and partial. There are numerous ways to classify nature and circumscribe the phenomena under study. Thus, humans are inevitably forced to make decisions on what to study, how to study, and so on. Human value judgments will dictate these decisions and, thus, determine how science is conducted and what knowledge is obtained. Furthermore, science is full of assumptions, concepts, and frameworks that are not themselves subject to empirical confirmation. This suggests that the very content of science is influenced by human preferences. In this regard, a distinctive contribution of Dupré is the idea that some concepts in science are themselves loaded with normative content (Dupré, 2007). This idea of 'thick' concepts can be illustrated with the example of inflation. While it might seem that inflation in economics is a topic that can be studied objectively without assuming human values, this is not the case. Decisions about how to measure inflation involve assumptions about what goods are valuable, how important different goods are in relation to one another, etc. These decisions reveal human preferences and, thus, show another dimension of complexity in science: acquiring knowledge about nature cannot be conducted from a value-neutral, view-from-nowhere perspective.

1.2 Genes and Microbes

Genomics occupies an important role in the development of Dupré's later philosophy. In the early 2000s, as the Human Genome Project was unfolding, genomics attracted a lot of interest from the humanities and the social sciences. Dupré was one of the main philosophers theorizing about the implications of those genomic projects. Around that time, he was involved in a large project on the social and philosophical aspects of genomics, namely the ESRC Centre for Genomics in Society at the University of Exeter. This project would ultimately expand in the form of the Centre for the Studies of Life Sciences (EGENIS, University of Exeter). From an intellectual perspective, thinking about genomics enabled Dupré to expand his view on the complexity of science and nature, eventually paving the way to his process ontology (Sect. 1.3). In this section, we start describing some of his contributions regarding the conceptual, social, and metaphysical aspects of genomics. Later, we turn to his philosophical views on microbiology.

The concept of the gene has attracted much philosophical interest (Fox-Keller, 2000). Dupré (2004, 2008) closely follows the development of genomics to argue that traditional concepts of the gene do not align with genomics science. On the one hand, the idea of Mendelian genes (i.e., independent heredity units with a determining effect in conspecifics) is largely inapplicable to contemporary genomics, as large portions of the genome are either non-functional or pleiotropic. On the other hand, the notion of Molecular genes (i.e., stretches of DNA) is too indeterminate, as it is unclear how to delimit genes. While some philosophers may attempt to either refine the molecular account of the gene or eliminate the concept, Dupré advocates for a "pluralistic" and "taxonomic" view of genes similar to his work on the species concept (Sect. 1.1). This view suggests that the term "gene" is primarily meant to create stability in classification, helping members of different biology communities to engage and connect their work. Thus, "gene" does not refer to any specific type of entity as dictated by molecular biology or any other area in biology. That term can legitimately stand for different kinds of material entities with no further unifying feature.

Dupré characterizes genomics as an important successor of traditional genetics. Genomics investigates a plurality of processes happening in the genome, abandoning the search for fundamental "laws of nature" that would describe the determinate effects of specific genes without considering environmental complexities. Genomics does not offer grounds for the commonsensical idea that there are "genes *for* X" ("X" being a determinate phenotype).[2] This conclusion has important social consequences. For instance, Dupré highlights that "there are no genes for race" (Dupré, 2008). Genomic technologies enable scientists to reliably infer the geographical ancestry of individuals, but this inference does not support the idea that races exist genetically in a meaningful way. There are no genes that make individuals part of one race versus another. Moreover, there is also no distinctive gene that exists in all and only

[2] This does not mean that genomics is not misinterpreted or presented to the public as a search for this.

members of a supposed race. Here, Dupré is one of the first philosophers to alert us to the risk of misinterpreting the science of genomics. Races exist as social kinds, largely as a social product of racism.

Genomics leads Dupré to increasingly appreciate life's diverse and dynamic nature. For instance, he challenges the commonsensical view that each organism corresponds to a single, distinctive genome present in all its cells (Dupré, 2010; Dupré & O'Malley, 2009). This view is falsified by numerous biological phenomena, such as mosaicism (mutations during cell division leading to cells with different genomes) and the microbiome (existence of bacterial genomes within multicellular organisms). These phenomena suggest that organisms and living systems are extremely complex, constantly changing, and often encompassing genetic material from multiple species. Collaborations exist within and between multicellular organisms, making it hard to draw clear boundaries between them. Furthermore, these phenomena carry lessons for the very notion of biological organism, namely: there is no simple, unique, and correct way to draw boundaries and thus individuate organisms. Ultimately, this view results in what Dupré calls "promiscuous individualism" (Dupré, 2012, 241): there is a level of arbitrariness to the boundaries we draw around organisms, and how we draw these boundaries can depend on the aspect of living systems we are most interested in. Again, Dupré finds a strong analogy between his earlier views on species and lessons from contemporary genomics.

One of Dupré's distinctive contributions to the philosophy of biology is a view concerning the nature of living systems. He and coauthors further explore the complexity and dynamicity of life by shifting their attention from multicellular organisms to *microbial* phenomena such as multispecies bacteria communities and symbiosis (Dupré & O'Malley, 2009). Relatedly, his philosophical analysis moves beyond genomics to metagenomics, i.e., the study of the DNA of microbial communities in their natural environments. The shift to microbiology and metagenomics grounds a new conception of living systems (Dupré & O'Malley, 2009). According to this view, there is a continuum between living and non-living entities with no sharp distinction between the extremes. Viruses, unicellular entities, eukaryotic cellular components, symbiotic entities, and multicellular organisms occupy different positions on the continuum. Dupré rejects the traditional criteria for considering those entities as living or non-living. For instance, while spatial boundaries are usually invoked as an important criterion, this criterion does not resist the complexity of microbial life and the arbitrariness of boundaries. Instead, the continuum between living and non-living represents the intersection between metabolism and lineage-forming entities. Life exists when lineage-forming entities collaborate in metabolic processes, which sometimes lead to different lineage-forming entities establishing multicellular entities or communities. Multicellular organisms like humans are extremely stable cooperative assemblages of lineage-forming entities (i.e., cells) connected through metabolism. This view of life puts cooperation and collaboration front and centre. It illustrates Dupré's insight that living entities are stabilized by processes such as metabolism. This insight is at the core of his process thinking, which we will turn to in the next section.

1.3 Process Thinking

Dupré's philosophy took a somewhat surprising turn in 2006 when he started to explicitly endorse a process ontology. It was surprising because Dupré now adopted a monism, something that seems at odds with his deep commitment to pluralism. In Sects. 1.3.1 and 1.3.2 we will trace the development of Dupré's process thought over the 15 years from 2006 to 2021. In Sect. 1.3.3 we will then analyse Dupré's mature process philosophy that emerged since 2022 and show how it helped him bring (philosophical) order to the disorder he diagnosed in his earlier work. We will also discuss why this turn to monism might not create a tension in his philosophy.

1.3.1 Beginnings

Dupré places the start of his process philosophy at a specific moment and place: 2006 in Amsterdam, where he was the visiting Spinoza Professor of Philosophy (Dupré, 2025a). More precisely, Dupré mentions a particular "Damascene moment" that (probably) triggered his conversion to process thinking, namely, a visit to the Albert Cuyp market in Amsterdam. According to his own recollection, he stumbled across this daily market by accident when exploring the city, as he was following a stream of people that caught his attention. As one might expect from a philosopher, he not only explored the market but also reflected on the nature of the phenomenon he witnessed. What is this object that he walked through? What are the features that make this phenomenon a "thing"? Dupré, swept up in the flow of people, concluded that it was this very flow of people and goods that made the market a persisting entity. The market *is* an activity, a confluence of different processes. It is not a fixed thing composed of well-defined parts.

In the Spinoza lectures that Dupré delivered that same year (published in 2008 as "The Constituents of Life"), he made first explicit references to the processual nature of biological entities, connecting his conversion moment to the entities he had been studying for decades. The backdrop to this initial engagement with process thought was not substance ontology—the traditional antagonist for process ontologists—but rather the machine-mechanistic vision of the world that, to this day, dominates much of the discourse in the life sciences and beyond. For Dupré, these mechanical models that assume "fixed machine-like ontologies" were abstractions from the dynamic and processual reality of biological systems (Dupré, 2008, 14). As he now saw it, life was not constituted by a hierarchy of well-defined machines and parts (or "things"), but by a hierarchy of dynamic and interacting processes (ibid., 12).

This beginning of Dupré's process thinking is instructive for at least two reasons: first, this switch in perspective allowed Dupré to make sense of the disorder, relativity, and lack of stability and clear boundaries that he observed in his work on biology. As we described above, he had long been pointing, for instance, to the fact that the genome is not simply a stable entity with well-defined parts and essential properties; it is a much more dynamic phenomenon with diffuse boundaries (Barnes & Dupré, 2009). An explicit process perspective allowed him to explain this and other biological phenomena.

Second, it is interesting that his move to a process ontology was not triggered by observing lineages, genomes or microbes. Dupré had been thinking about these entities for two decades, but he was prepared to accept the disorder he encountered in this realm, and in science more generally, as a brute fact (Dupré, 2024). What triggered the move to an explicit process thinking was a socioeconomic phenomenon. This is relevant because, as we will see in Sect. 1.3.3, Dupré's thinking is always also about the human and the social. Science matters because it shapes how we talk about "the" human, or gender, race, intelligence, violence, and so on. But what is at the heart of Dupré's work on biology is the desire to re-think what science is telling us about these concepts, entities, and processes, and to thereby challenge some of the dominant narratives and assumptions about humans and the things they do or should do.

After the 2006 turning point, Dupré's process view gradually developed over the course of 20 years. For ease of discussion, we will distinguish between an early (2006–2021) and a mature phase (2022–present) of his process thinking. In the early phase we not only observe the gradual emergence of the key ontological tenets that Dupré subscribes to, but also a detailed explanation of how they are supported by scientific findings and how they help inform or resolve ongoing debates in science and philosophy. In his mature phase, which will be the focus of Sect. 1.3.3, we find a refinement of his central ontological tenets and, importantly, a broader application of his process thought to issues such as personhood and free will.

1.3.2 Early Phase

The early phase of Dupré's process thought was characterised by two key aims: first, to show that we need to engage more explicitly with ontology if we want to model and understand biological systems. As he put it in his Spinoza lectures:

> […] what we need as a theoretical infrastructure for modelling biological systems […] is ontology. We cannot expect to understand the behaviour of molecules in a cell unless we have a clear idea of what sorts of molecular objects there are […]. (Dupré, 2008, 48)

This explicit focus on building an ontological framework was an important departure from his previous work, as even though he engaged with metaphysical issues throughout his career, such as the question of natural kinds, he never aimed to build or endorse a specific ontology. The move to a process ontology thus added a new element to Dupré's work.

The second aim was to establish a research agenda that could help achieve the first aim. An early outline of this research agenda can be found in the paper "Living

Causes" (Dupré, 2013) where Dupré presents three pillars of his emerging process thinking:

1) The world is processual.
2) Stability is relative.
3) Ignoring 1) and 2) can have problematic epistemological consequences.

These three claims bring together ontology, methodology and epistemology. The first claim establishes the basic picture of the world: everything is a process, and change is therefore the default state of the world.[3] The second claim is an ontological claim that contains a methodological directive: we need to constantly adjust and reflect on the time scales we apply when analysing an object. This will help us show that the stable things we can observe in nature are in fact just processes stabilised at different time scales. A mountain may be a stable thing to the hiker walking up it. But if we observe it over hundreds of years, we will see that is undergoing constant change. This helps us show that the stabilities we observe in nature are relative to the time scale the observer applies; they are relatively stable processes. Following the methodological directive included in claim 2) thus helps support claim 1). The third claim is epistemological in nature: it warns us of the (epistemic) dangers of not applying 2) and of ignoring 1). The main issue is that we might draw flawed conclusions when studying the world because we are applying a flawed ontology.[4]

Much of the work in the early phase of Dupré's process thought was dedicated to teasing out the implications of adopting—or neglecting—claims 1) and 2), thereby elaborating on claim 3). In the following we will summarise some of the key issues and facts Dupré engages with, starting with the issue of persistence.

The question of persistence connects to process thought in at least two ways: first, a process ontology seems to capture better what science tells us about the persistence of biological entities. This means a process view is empirically adequate in terms of the *content* of science (Dupré will also argue that it better captures what scientists are actually doing, see below). Second, Dupré argues that a process ontology allows us to avoid some of the problems that a substance or thing ontology faces when dealing with the question of persistence, meaning it is philosophically more fruitful.

---

[3] Note that in his early phase Dupré often qualifies this claim, restricting it to biological entities or living systems. For instance, in a 2015 article Dupré states: "I want to claim that an ontology of processes is far better suited to understanding the nature of life and the living than the more standard ontology of things" (Dupré, 2015, 33). And Dupré and Nicholson write in the introduction to the book "Everything flows": "The particular metaphysical thesis that motivates this book is that the world—*at least insofar as living beings are concerned*—is made up not of substantial particles or things […] but of processes" (Nicholson & Dupré, 2018, 3, emphasis added). They further qualify this claim by adding that "The essays in this volume offer many reasons for thinking that the living world is a world of processes rather than a world of things, although we do not take this to demonstrate conclusively that the world is processual throughout" (ibid., 4). They add that there are "compelling reasons" to assume that the world is processual throughout, but they remain cautious. As we will see in Sect. 1.3.3, in his mature process phase Dupré drops this caution and defends a universal processualism.

[4] These three claims were further elaborated on in the 2018 "process manifesto" that Dupré co-authored with Dan Nicholson, and which served as the introduction to the edited volume "Everything Flows" (Nicholson & Dupré, 2018).

We will first address the question of why a process view is empirically adequate. Dupré goes back to the basic claim that a process *is* activity or flow. A process is change, and if the activity stops, the process has stopped existing; a process requires change for its continued existence (Dupré, 2025b). Incidentally, this is also the case for an organism: for an organism to persist it relies on a range of processes, such as a beating heart or a set of moving limbs. But most importantly perhaps, all organisms depend on thousands of metabolic reactions that have to run in a highly integrated manner to make sure that the organism does not perish. As Dupré put it in "Living Causes", a mouse that is not active is an ex-mouse (Dupré, 2013, 31). Metabolic activity also means that there is a constant turnover of the matter that makes up the organism at any point in time and at every level of organisation. Organisms are "metabolic streams of matter and energy that exhibit dynamic stabilities relative to particular timescales" (Nicholson & Dupré, 2018, 17).

The dynamicity of organisms is not only expressed in their metabolic activity but also in the life cycles they undergo; they develop and change their morphology and behaviour over the course of their life. But this fact is easily overlooked if we ignore claim 2) and use a narrow time scale to reflect on the nature of organisms. This often leads us to focus on the adult stage whilst ignoring other stages that form part of the larger life cycle of an organism. This simplification can be problematic because the adult stage is usually a relatively stable stage that most resembles a "thing". This can easily lure people into treating the organism as a thing (Nicholson & Dupré, 2018, 19).

But the adult stage is not "the" organism. An organism such as a frog is the whole cycle, which includes the egg or the tadpole that later develop into an adult frog. As Dupré highlights in his later work, it is not the case that one being comes to an end (the tadpole) and a new comes into existence (adult frog) when we move along the life cycle of the frog; they are just radically different stages of the same organism or process (Dupré, 2025a, 9).

The different stages of this process don't share any obvious properties between them, as they differ both structurally and behaviourally. There don't seem to be any essential properties of the life cycle. The main candidate for continuity through the cycle would probably be the genome, but it is not clear what would qualify as "the" genome of an organism, due to the highly dynamic nature of genomes and the polygenomic nature of organisms (Dupré, 2010).

The above not only applies to organisms such as humans or butterflies, but also to viruses. Even though viruses are usually represented as a particle or "virion"—a three-dimensional structure consisting of viral (and sometimes host-derived) proteins and genetic material (DNA or RNA)—this "thing" is only one stage of the larger process that is the virus (Dupré & Guttinger, 2016). For a significant part of a viral cycle there is in fact no particle: the virion is disassembled when it infects a cell, leaving behind just the viral genome and some dispersed viral proteins inside an infected cell; the virion has ceased to exist. But surely we would not claim that the virus has ceased to exist simply because we have moved past the cell entry phase. The genomic material then co-produces with the cellular infrastructure more viral proteins and DNA or RNA, which are then assembled into new virions. These particles exit the cell, ready to infect more cells. This cycle is what we refer to when we say that "a virus" has infected us. Importantly, there is again no single property that unifies all these viral stages, as even the genetic material of the new particles is often highly mutated or re-mixed compared to what originally entered the cell. The virus is a *process*, and not a discrete, autonomous, and stable thing (ibid.).

As the above shows, the existence of life cycles helps illustrate the absence of essences (Dupré, 2018a) as well as the fundamental dynamicity of living systems. But it also provides support for the claim that a process ontology is philosophically more fruitful than a thing or substance view, i.e., that it can deal with problems that a thing view cannot deal with. A thing view of the world assumes that there are well-defined discrete things with a given set of properties. On this view, two things are identical if they have the same properties. But if a human, for instance, has blonde hair when they are a baby and brown hair when they are an adult, then they don't share the same properties and would, in a simple thing view, not qualify as the same entity. The problem for a thing ontology is thus to explain how an entity can change whilst remaining the same thing (Dupré, 2018a). This problem becomes even more challenging when we turn our attention to entities with a life cycle, as the level of difference in terms of physiology, morphology, or also behaviour goes well beyond a simple difference in a single trait such as hair colour.

For a process ontology, however, the example of biological entities and their life cycles does not pose the same sort of challenge. Processes have temporal stages, and being part of the same process is not about the degree of similarity between different stages but about having a continuous and causal connection to earlier stages (Dupré, 2018a). A process view can thus solve issues that a thing view struggles to overcome.[5]

There is also an epistemological-methodological benefit to a process view: it captures more precisely what scientists are actually doing when they build new knowledge. In a thing view, stability is the default state in nature. What needs explaining is how stable, well-defined things change from one state to another.

But the key questions that scientists address are how nature *establishes* stable structures or patterns, and what role these stabilities play in the process of life. For instance, looking at the discipline of physiology and its focus on structure and function of the body, Dupré states that:

> Physiology does not investigate the properties of a stable object, but the processes that enable a system of some kind to retain its form sufficiently for it to continue to function. (Dupré, 2015, 34)

A process view can make sense of this, because it sees change as the default state of the world. What needs explaining is stability, i.e., how a confluence of different processes can bring about temporarily stabilised islands in a sea of change. This points to a further benefit of adopting a process view: it is not only empirically adequate in the sense that it captures the nature of biological entities, but it also allows us to capture what scientists are actually doing and saying.

---

[5] This of course does not mean that a process view will not run into problems of its own. Dupré is aware of this, and we will comment in Sect. 1.3.3 on how he proposes to deal with at least some of them.

Adopting a process ontology also allows us to provide new answers to old questions. An example is the discussion about the nature of living systems. Most traditional accounts put metabolism and reproduction at the heart of their accounts: what marks the difference between a living system and, say, a genome is that the former engages in metabolism and reproduces itself.

But such assumptions of autonomy are exactly what Dupré, using a process-inspired lens, would question. As we now know from various studies on the symbiotic nature of living systems, even multicellular organisms cannot perform necessary processes such as metabolism without collaborating with other biological entities. Bacteria take over key functions in the digestive process, and even viruses can be seen as vital functional parts of the whole (Dupré & Guttinger, 2016). Specific examples include cyanophages, the viruses that infect cyanobacteria: these viruses have been shown to actively modulate the bacterial metabolism, making them a key player in this process (Ain et al., 2024). Traditional accounts thus miss a central feature of life, namely the activity of "collaboration".

This has important implications for how we draw boundaries around living systems: a perspective that treats living systems as a series of dynamic collaborations between different lineage-forming entities emphasises that the boundaries of living systems are not fixed (Dupré & O'Malley, 2009; see also Sect. 1.2). A virus, for instance, could be treated as being alive whilst it is actively collaborating with another entity and lose this characteristic when it leaves this collaboration (ibid.). What counts as alive is thus not always clear and stable—the boundaries of living things are fuzzy and change over time.

A thing ontology struggles to account for such fuzzy boundaries, as it treats entities as being independent of the various relations they enter. The relations are always external to the thing, which assumes a clear boundary between the discrete thing and its surroundings (Nicholson & Dupré, 2018). This thing view includes the idea that all the capacities and key properties belong somehow to the inside of the thing; it carries these properties independently of the larger context. But as Dupré and Nicholson highlight, for a process ontologist the interconnectedness of life challenges these assumptions. The boundaries of an organism are vague and what are often called "external" factors can be part of what endows the organism with certain properties or capacities (ibid., 21).

These insights connect to an important aspect of Dupré's earlier work. The process view allows us to explain why we should be pluralists when it comes to classification (Dupré, 1993a) (see Sect. 1.1). It also shows us why we should endorse "promiscuous individualism" (Dupré, 2012, 241). How we draw these boundaries can depend on the aspect of living systems we are most interested in. The processes of metabolism will point us to different relevant boundaries than what immunological processes would guide us to. Other players and/or relations will become central to the relevant functional unit carrying out these processes, depending on what element of life we aim to study or intervene in.

This can also be seen when we consider the tree of life, a central concept in evolutionary science. As Dupré argues in a paper co-authored with Eric Bapteste, the traditional tree of life applies a narrow lens to the process of evolution, as it only focuses on vertical inheritance. Each branch of the tree represents a reproductively linked sequence of entities, derived from an ancestral entity. Importantly, the flow of genes is constrained to this branch, as they are seen by the "standard evolutionary ontology of biology" as passengers of the organisms that reproduce (Bapteste & Dupré, 2013). The key issue here is the strictly hierarchical and parts-based view applied to living systems by this standard ontology.

Yet science tells us that genes also undergo horizontal transfer—they move between organisms that occupy different branches of the tree. This means that the genetic material we find in an organism does not necessarily have its origin in the parent organism. It could have been brought in, for instance, by a virus, or it could be transmitted from one bacterium to another, even when they don't form a parent-offspring pair. These and other phenomena mean that the traditional tree picture is in trouble.

The problem, according to Dupré and Bapteste, is that the standard ontology privileges only one type of relation (vertical inheritance) and entity (parent-offspring pairs). This not only neglects other sources and relations (e.g., horizontal gene transfer) but it also ignores the different time scales at which the various entities, or processes, are stabilised. The tree of genes, for instance, will not look the same as the tree of organisms, as these are different kinds of processes that are stabilised at different timescales. Tracing a set of relatively stable genes (in evolutionary terms), such as the ribosomal genes, might not tell us much about the stability of the organisms that carry them in their lineage. With its hierarchical picture of living entities, the standard ontology simply assumes a whole-part coupling that is, in light of scientific findings, not warranted (ibid., 389).

1.3.3 A More Refined Process View

In the mature phase of Dupré's process thought we encounter a refined ontology. This refinement is achieved in two steps: first, there is a clearer articulation of what a "thing" is. Whereas in his early process phase the alternative worldview was more vaguely framed as either a "machine-like" ontology or a substance or thing ontology, the details of what a thing view exactly entails were not worked out systematically. Building on this sharper picture of things, we then also get a more developed positive account of what a process is.

Dupré identifies five key attributes of "things":

1. They are stable by default.
2. They don't need to undergo change in order to persist.
3. They have essential properties that determine what kind of thing the thing is.
4. They exist autonomously, meaning they don't depend on other things to exist.
5. They have clear and well-defined boundaries; they are discrete entities.

As we have seen above, Dupré has aimed to show that none these characteristics apply to biological entities. Organisms are dynamic rather than stable (metabolism, heartbeat, collaborative agency etc.); they need to undergo constant change in order to persist (metabolic activity); there are no clear essential properties we could ascribe to them (evolution, development, life cycle); they don't exist autonomously (symbiosis, collaboration); and they don't have well-defined boundaries (symbiosis).

This establishes that an organism is not a thing, as defined by Dupré. But this presents us with a negative characterisation of organisms: it tells us mainly what organisms are not. It does not tell us what a process is and what it means to say that an organism *is* a process.

In his mature phase, Dupré starts to develop a more detailed view of how we should understand processes, beyond claim 1) of his early process phase (i.e., "The world is processual"). First, he defines a process as something that requires change for its continued existence (Dupré, 2025a, 2025c). He thus postulates a necessary (or essential?) condition for something to be a process.

Second, Dupré taps into the traditional distinction between "occurrents" and "continuants" to further elaborate on what he thinks processes (and things) are. Occurrents are entities that have temporal parts; they are "in time" (Simons, 2018). Examples include events, such as conferences, or processes, such as continental drift (Simons, 2000). Continuants, on the other hand, are entities that don't have temporal parts, they are numerically one and the same at different times—my car or my cat is the same it was yesterday (ibid., 61). These "continuants" are what philosophers usually refer to as "substances" (Simons, 2018). In most general terms, they are the three-dimensional objects we interact with in everyday life (Stout, 2016).

Dupré rejects this way of dividing the world. A key issue is that authors such as Simons use organisms and rivers as examples of continuants (substances). To Dupré, however, these phenomena are paradigmatic examples of processes. Following (Stout, 2016), Dupré now claims that they should be seen as "continuant processes".[6] With this move, Dupré creates two subclasses of processes: some processes can be seen as continuants, and others as occurrents. But all are processes.

Dupré argues that Simons misses this point because he does not distinguish carefully enough between how we speak about the world and how the world is. These are two different issues, but Simons seems to confuse the two. Dupré agrees with Simons that substances (continuants) are entities of which we *say* that they remain identical over time. But this does not mean that they in fact remain the same. These continuants share features of processes, no matter how we tend to speak about them (a car, a person, etc.).[7] They are thus processual continuants (Dupré, 2024).

[6] Note that Stout speaks of "occurrent continuants" (Stout, 2016). His aim is to show that the two categories are not mutually exclusive, meaning that it is possible to postulate that there are substances or three-dimensional physical objects that have temporal parts, and events (occurrents) that have no duration such as, for instance, an accomplishment like "reaching the top of a mountain" (ibid., 42). But beyond that, he wants to argue that some processes have a duration but don't have temporal parts—they are occurrent continuants or "dynamic" continuants (ibid., 60).

[7] It is not clear whether Simons would disagree much with Dupré's take on processes and substances. Simons holds (or at least does not reject) the idea that processes (occurrents) are metaphysically more fundamental (or "basic") than substances (continuants) (Simons, 2000, 75). He, for instance, postulates that there are "vital occurrents", which are the processes required for continuants to

1.3.4 Tackling Old and New Challenges

As we already pointed out at the beginning of Sect. 1.3, adopting a process view allows Dupré to further support some of the claims he has defended throughout his career, including his rejection of essentialism. A process view gave Dupré a new tool to show that people pay too much attention to prominent phenomena or states, taking the part to stand for the whole. This already came up in his work on determinism, essentialism, and reductionism: the order we observe, and which is so important for these positions, is always only local. People, ultimately, read too much into examples of (local) order. A process view, in combination with the methodological directive entailed in claim 2 (Sect. 1.3.2), allows Dupré to explain in more detail when and how people pay too much attention to isolated parts. Rather than having to point to the local or the niche, he can now point to the whole: people should consider, for instance, the *whole* life cycle of an organism, and not just the adult stage. A process view allowed Dupré to make a positive rather than a negative case for local order. Instead of just saying "there is no universal order", he can say that a unified universe explains why regularity and stability are always local and why we have a disunified science (Dupré, 2024).

In his mature phase Dupré also uses the more refined picture of a "thing", and his process view more generally, to re-think other issues that have been of interest to him throughout his career, such as the nature of causation in biology, the question of personhood and personal identity, or the problems of human nature and free will. We'll pick the question of human nature here to illustrate how Dupré's view of "things" and processes shaped his approach to these larger philosophical, social and historical issues. The main target of his critique of human nature is the view presented by researchers in fields such as sociobiology and "Evolutionary Psychology" (Dupré, 2012, ch. 16). In simple terms, this school of thinking attempts to explain human behaviour by reference to internal psychological modules and mechanisms that were shaped by evolutionary processes; there is a particular cognitive (brain) architecture, and this architecture is an adaptation to the pressures our ancestors have been exposed to. How we react to external and internal stimuli is shaped or even determined by mechanisms that our ancient ancestors acquired when trying to survive and reproduce in the Pleistocene (ranging from about two million years ago to about 12,000 years ago). As a result, there is a "human nature" that is common across time and members of the species.

In his early critique of these views, Dupré's anti-reductionism took centre stage. What was wrong with approaches like the above was to assume that our mind can be reduced to the brain, and that this brain is some sort of machine that can be reduced to a set of genes and modules. In epistemic terms, this means that to understand and predict human behaviour we simply need to know the genes and the modules that shape our mind and behaviour. And methodologically the research on human nature becomes a hunt for the genes that cause certain types of behaviour or human traits.

---

persist. An example are rocks (continuants) which are "teeming with occurrents [processes]", in particular the movement of electrons that allows a stone to continue to exist (ibid., 70).

In Dupré's view, this framework is too limited to even start to make sense of the diversity and plasticity that human behaviour displays. In part, this has to do with the range of factors that shape our behaviour: what we do, or will do, does not just depend on specific genes and modules, but also, for instance, on the opinions and behaviours of other humans, and on how we develop through our interactions with our environment (Dupré, 1999). Humans are of course not just blank slates—they are part of a biological lineage. But there is a complex landscape of local factors, rather than just a set of genes handed down in a linear parent-offspring chain, that explains the large variety in human behaviour we observe in everyday life.

In his later process phase, this anti-reductionist foundation is still present, but Dupré's view of human nature now focuses more on the idea of the human as a process. Developmental dynamics, life cycles, symbiosis, epigenetics, niche construction, or the process of learning—all of these processes are part of the stabilised yet plastic process that a human is. This puts the focus not just on the diversity of factors that shape human behaviour but also the temporal dimension: humans can quickly adapt to local circumstance, for instance through adjustments to the composition and activity of their microbiome (symbiosis and metabolism), or through epigenetic changes that don't rely on genetic mutations to take hold in a population, or through cognitive processes such as learning. Humans are not mere "things" with a given set of properties, but open-ended developmental processes that are temporarily stabilised by a multitude of processes and activities (Dupré, 2018b). This also means that differences in how humans behave are not simply due to mere genetic differences, but to their ability to quickly adjust to different contexts (Dupré, 2025a).

If by "human nature" we mean something like the things that humans do in a specific place and time, then Dupré has no quarrel with this concept. But from such a minimal local history we cannot easily generalise, and there is no straightforward way of predicting how a human will behave; humans are not machines that simply run on a pre-set programme. In Dupré's view, there is thus no point in trying to define or identify something we could call "the" human nature that applies to all humans across time and space (Dupré, 2018b).

This is just one example of how Dupré has further expanded on his earlier work, using the process lens to show why the thing-focused approaches he opposed throughout his career present a less convincing and accurate view of the living world.

1.4 Interdisciplinarity and Engagement

Let us now consider Dupré's style and method of philosophical scholarship. It is important to highlight that Dupré accomplished all of the above in constant dialogue and frequent collaboration with practicing biologists as well as historians, sociologists, anthropologists, geographers, and metaphysicians. Collaboration with biologists is by no means exceptional within Anglo-American philosophy of biology, and yet the breadth and depth of Dupré's own engagement is impressive and highly significant for the development of his philosophical ideas. It is clear, for instance, how an in-depth study of taxonomy—and particularly botany—inspired Duprè's early work on promiscuous realism and anti-essentialism concerning natural kinds. The influence of biologists—as well as historian of biology Nick Jardine, a specialist in the history of natural history who was Dupré's PhD supervisor at the University of Cambridge—impressed upon the young Dupré an awareness of how biological inquiry not only reflects, but thrives in consideration of the endless variety of forms which Darwin himself so aptly attributed to organic phenomena. Similarly, Dupré's later anti-reductionist stance was intertwined with his study of—and lively debates with—genomics at a crucial phase of its development as a scientific field, with technologies such as genetic sequencing coming of age as mainstream forms of data production and intervention for the life sciences, with related worries about the potential to overemphasise molecular structures over the physiology and development of cells, tissues, organs, and ecological relations among organisms. And the processual philosophy developed in later years matches both the closer attention paid by some biologists to such complex ecologies and interactions, and the dangers created by the increasing reification of biochemical and molecular mechanisms as the lens through which contemporary life sciences are financed and used to feed societal technologies and imaginaries of the future (through agricultural interventions such as the implementation of genetically modified or otherwise engineered organisms, for instance, or highly gene-centred forms of precision medicine).

This brings us to the second characteristic of Dupré's work we wish to highlight in this introduction, namely, his commitment to bringing social concerns around the implications of contemporary research developments in biology into the philosophy of science, and to do so in a way that is both empirically informed and normatively charged. This commitment is evident in his acclaimed work on science and values and particularly the ways in which biological research embodies specific—and sometimes highly questionable—understandings of what it is to be human and a social animal. This body of work not only attacked value-free understandings of science but also affirmed the responsibility for philosophers of science to take a critical stance on the role and implications of research in society. Fulfilling this responsibility, in Dupré's career, has meant engaging in productive collaborations not just with the biologists whose work he wanted to understand and sometimes contribute to, but also with social scientists and historians from around the world whose methods inform an appropriate contextualization of biological research in relation to its social, political and economic settings—as well as policy-makers, public intellectuals and journalists working to use research outputs as evidence for societal decision-making. One would imagine that such a collaborative, inclusive approach may stumble upon interdisciplinary and transdisciplinary conflicts, vehement disagreements and institutional obstacles, and in working with him we have certainly witnessed those arise: but Dupré's particular brand of optimism, combined with his ability to listen to others with an open mind, to gracefully and constructively navigate complex social interactions, and to always come up with the right joke at the right time, always seems to win the day. This brand of engaged philosophy of science has also paved the way for the approach now known as 'philosophy of science in practice' (Ankeny et al., 2011), which Dupré has supported and fostered from its inception, and which is now busy formalizing what this vision of philosophical scholarship means in methodological terms (Veigl & Currie, 2025).

The unparalleled effectiveness of Dupré's engagement with a variety of colleagues and relevant peers across disciplines and contexts, and the gift that this continues to be to our profession, is wonderfully exemplified by Egenis, the Centre that John created in Exeter in 2002 and through which for the past twenty years he has been able to host a constant stream of philosophers, sociologists, historians, and scientists from all walks of professional life and sometimes radically different persuasions. It is one thing to conceive of a philosophy of science that is congenial to the concerns of working biologists, STS scholars and other social scientists, and historians of biology; it is quite another to create and foster an institution that embodies and nurtures robust bridges between disciplines, ideas, and people. It is thanks to Dupré that several interdisciplinary-minded and socially engaged philosophers of the life sciences, including us as the editors of this volume, have found a welcoming home and a lively and continuously inspiring and supportive community in Exeter.

Leading by example, Dupré has long complemented his research activities with a keen involvement in public engagement. He has encouraged and participated in public debates with a wide variety of non-academic audiences throughout his career and regularly intervenes in policy debates around genetic determinism and its biomedical and wider social implications. He has also played significant roles in public debates and policy consultations over evolutionary psychology, neo-Darwinism and, most recently, gene editing. For instance, Dupré has chaired the Nuffield Foundation Committee on gene editing of livestock, a task that saw him work with policy-makers, media, food industry representatives and lobby groups involved in this area in the UK. Dupré's work across all these contexts has always been driven by strong social concerns and a commitment to articulating the values that underpin contemporary research and debating what directions research ought to take in order to serve the long-term flourishing of humans and non-humans on the planet. In his everyday activities, philosophical ideas have consistently flown into social interventions, and vice versa—a feat that includes his relationship to colleagues and students.

This takes us to the last characteristic of Dupré's scholarship that we want to highlight: his immense humanity, his generosity and thoughtfulness as a mentor and teacher, and his tireless leadership of our field, exemplified not only by his long-standing service to the Philosophy of Science Association (PSA), but also to the British Society for the Philosophy of Science, the International Society for the History, Philosophy and Social Studies of Biology, and many other scholarly bodies. As we witnessed Dupré serving in his many institutional roles, we have never seen him brag about his work, act condescendingly to others, or impose decisions without real consideration of other viewpoints. Rather, we continue to learn from his understanding of what inclusivity means in practice, his unceasing intellectual curiosity and his inexhaustible willingness to mentor numerous students and advise junior faculty. No matter the requirements on his time, his door has always been unfailingly open to colleagues and students from around the world—which of course means that his work has touched and often transformed the lives and careers of hundreds of scholars, including our own.

What we hope to leave readers with is a sense of the many ways in which Dupré's work exemplifies the pursuit of philosophical insight that is grounded in rigorous and critical understanding of scientific knowledge and practice, open to collaboration with and insights from both the natural and the social sciences, mindful of the urgent need to increase diversity and equality within and beyond our field, passionate about research, education and societal engagement, and full of wit, care and love for life.

1.5 Overview of Chapters

The previous sections have provided an overview of Dupré's impact on philosophy. They covered central themes, arguments, and methodologies developed by John Dupré over the last 45 years. While this review is far from complete, it offers the reader a basic background to the chapters of this volume. These chapters engage critically with Dupré's work, sometimes expanding, sometimes rejecting his views, sometimes doing both at once. In their own way, each chapter advances debates in philosophy by exploring core themes of Dupré's philosophy in light of contemporary science. The result is a series of original contributions borne out of thinking *with* and *beyond* Dupré's work. These contributions are organized in three parts, each of them containing chapters centred around one of the following aspects of Dupré's work: disorder of science, process ontology, and interdisciplinarity. Disorder and process constitute the two central themes of his philosophy, while interdisciplinarity captures the core methodological commitment underlying them. This organization is somewhat artificial, however. As the reader will notice, many chapters cut across the different themes and methodological approaches of Dupré's work.

In Part I, chapters mainly engage with themes of Dupré's work. Catherine Kendig (Chap. 2) explores the themes of classification, natural kinds, and processes to investigate the nature of soil, an extremely undertheorized topic. She investigates how soil testing—from the early use of soil fertility trains that brought soil lab testing to farmers in the 1920s to more recent "tighty whitey" tests of soil biological life rely on processual commitments about soil. Soil is not just inert dirt. It includes the communication of root exudates and an extended biochemically facilitated microbial community. Assessing the health of soil means assessing soil microbiological activity. Kendig demonstrates how a Dupré-inspired analysis provides tools to uncover implicit and explicit use of the ontological commitments, kinds, and models that are relied upon to make sense of these soil assessments. In turn, Nancy Cartwright (Chap. 3) builds on the themes of complexity, anti-reductionism, and causation in the sciences. Cartwright elaborates on Dupré's idea that economics can only provide "loose knowledge," arguing that economics should forego its pride in being quantitative and exact. Instead, economics should export her rigorously-established principles as *qualitative, inexact tendency principles*. Though inexact once exported to the large complex world, the rigorous backing these principles receive in economics' small-world models can be helpful in figuring when we can expect these tendency principles to obtain. Finally, Adrian Haddock (Chap. 4) identifies and examines a fundamental tension between anti-reductionism and empiricism within Dupré's philosophy. The chapter consider John's resistance to the idea that the ultimate ground for rejecting reductionism is the subjectivity of the mental. John rightly recoils from traditional arguments against reductionism in the philosophy of mind, but he does not consider arguments which understand subjectivity as self-consciousness: a form of cognition which is at once first personal and non-empirical. Haddock suggests that John should endorse the idea in this version. Nevertheless, Haddock also suggests that he cannot do so because, if he were to do so, then he would no longer be an empiricist, and empiricism is integral to John's philosophical identity.

In Part II, the chapters revolve around process philosophy. Papale, Doolittle and Inkpen (Chap. 6) examine process thinking in the context of natural selection, showing how holobiont and multispecies assemblages can be best understood as processes in that context. Processualism allows the authors to strengthen their famous, preferred characterization of natural selection (ITSNTS—"It's the Song, not the Singers"). At the same time, the authors expand on process philosophy as a view about explanation rather than ontology, moving away from Dupré's metaphysical leanings. Scott Gilbert (Chap. 5) moves beyond the evolutionary context, expanding and grounding processualism scientifically. For him, multispecies interactions are fundamental for the three main processes of life: metabolism, development, and evolution. Recent studies show that what we had thought of as metabolism based on a single genome is actually the co-metabolism of more than one species. What we had thought as development from a single zygote is actually sympoiesis, co-development, where symbionts provide important signals for normal cell differentiation and maturation. And what we had thought of as evolution is the co-evolution of several species together. In Chap. 8, Tim Lewens examines human nature through the lens of processualism. Human Nature is a concept that has considerable attractions. It is also one that presents considerable problems. John Dupré has argued that a process-based approach leads to a better understanding of why the Human Nature concept has such ambiguous appeal. This chapter follows Dupré's lead, using the disordered plurality of processes that underpin species-wide generalisations, and species-wide stability, to argue for a sceptical approach to the very idea of Human Nature. Thomas Pradeu (Chap. 7) returns to the topic of processes and biological interactions, focusing on symbiosis. He shows how symbiotic interactions between 'macrobes' (non-microbial organisms) and microbes, is a central topic in John Dupré's work. He spells out why symbiosis has turned out to be a sort of ideal trope for Dupré, as it encapsulates many of his main personal themes—from promiscuous realism to process ontology. Then, Pradeu shows that immunology offers one interesting and complementary entry point to those themes, considering issues around microbial symbiosis and biological individuality.

In Part III, chapters reflect on interdisciplinary approaches to philosophy, a key methodological orientation of Dupré's work. Jane Calvert (Chap. 9) start asking what interdisciplinarity means in the context of studies of the life science, then moving to a more intractable question: could we even think of collaborating with other organisms, generating studies *with* rather than *of* them? Calvert end by arguing that opening ourselves up to being transformed by our objects of study will help us build new solidarities across disciplines, which are becoming increasingly important as we face planetary-scale changes. Next, Jim Griesemer (Chap. 10) challenges some of the core topics in Dupré's theory and practice. Fors instance, there is the questions of how to be a pluralist: whether to be a monist of the moment and choose an ontology that is better or best for the problem at hand in that moment's context, or to be a multi-perspectivist, i.e. to seek understanding in alternative *complementing* "ontologies" at the same time as aids to navigating complex phenomena. Griesemer defends the latter, while Dupré seems attached to the former. Additionally, representationalist views of knowledge seem to be a reflection of epistemology/metaphysics dualisms. While Dupré may still operate within this framework, Griesemer challenges it with a Wimsatt-style multi-perspectivism. These challenges echo a radical need for interdisciplinarity, one that may go beyond how Dupré practices it. Finally, Alan Love (Chap. 11) examines more concrete aspects of interdisciplinary collaboration based on his own research. First, he examines a large-scale interdisciplinary project that interrogates one of the most controversial aspects of life science: teleology. Second, he discusses interdisciplinarity at a more local level, considering how interdisciplinarity shapes the activities at the Minnesota Center for Philosophy of Science (University of Minnesotta). These examples of institutionalized collaborative efforts illustrate the distinctive value of jointly exploring the assumptions and evidence behind scientific models to advance our understanding of how the sciences work and what they imply.

The volume closes with two additional chapters. In Chap. 12, Phillip Kitcher embraces a more personal and poetic tone to reconstruct Dupré's contributions to philosophy, while also criticizing one of his key tenets: promiscuous realism. Kitcher points out the need for ethical considerations when doing metaphysics, inviting Dupré (and fellow pluralists) to embrace pragmatism rather than realism. In the final Chap. 13, John Dupré addresses this latter point, as well as some of the issues raised in the previous chapters of this volume.

Acknowledgments ERC-STG-2024 number 101161994.

References

Ain, Q. U., Wu, K., Wu, X., Bai, Q., Li, Q., Zhou, C. Z., & Wu, Q. (2024). Cyanophage-encoded auxiliary metabolic genes in modulating cyanobacterial metabolism and algal bloom dynamics. *Frontiers in Virology, 4*, 1461375.

Ankeny, R., Chang, H., Boumans, M., & Boon, M. (2011). Introduction: Philosophy of science in practice. *European Journal for Philosophy of Science, 1*(3), 303.

Bapteste, E., & Dupré, J. (2013). Towards a processual microbial ontology. *Biology & Philosophy, 28*(2), 379–404.

Barnes, B., & Dupré, J. (2009). *Genomes and what to make of them.* University of Chicago Press.

Bird, A., & Tobin, E. (2024). Natural kinds. In E. N. Zalta & U. Nodelman (Eds.), *The Stanford encyclopedia of philosophy* (Spring 2024 ed.). https://plato.stanford.edu/archives/spr2024/entries/natural-kinds/
Cartwright, N. (1983). *How the laws of physics lie*. Oxford University Press.
Coyne, J. A., & Orr, H. A. (2004). Speciation: A catalogue and critique of species concepts. In *Philosophy of biology: An anthology* (pp. 272–292). Wiley.
Dupré, J. (1981). Natural kinds and biological taxa. *The Philosophical Review, 90*(1), 66–90.
Dupré, J. (1986). Sex, gender, and essence. *Midwest Studies in Philosophy, 11*(1), 441–457.
Dupré, J. (1993a). *The disorder of things: Metaphysical foundations of the disunity of science*. Harvard University Press.
Dupré, J. (1993b). Could there be a science of economics? *Midwest Studies in Philosophy, 18*(1), 363–378.
Dupré, J. (1996). Promiscuous realism: Reply to Wilson. *The British Journal for the Philosophy of Science, 47*(3), 441–444.
Dupré, J. (1999). On the impossibility of a monistic account of species. In *Species: New interdisciplinary essays* (pp. 3–22). MIT Press.
Dupré, J. (2001). In defence of classification. *Studies in History and Philosophy of Science Part C: Studies in History and Philosophy of Biological and Biomedical Sciences, 32*(2), 203–219.
Dupré, J. (2002). *Humans and other animals*. Oxford University Press.
Dupré, J. (2004). Understanding contemporary genomics. *Perspectives on Science, 12*(3), 320–338.
Dupré, J. (2007). Fact and value. In H. Kincaid, J. Dupré & A. Wylie (Eds.), *Value-Free Science? Ideals and Illusions*. Oxford University Press.
Dupré, J. (2008). *The constituents of life*. Koninklijke Van Gorcum. (Reprinted in Dupré 2012.)
Dupré, J. (2010). The polygenomic organism. *The Sociological Review, 58*(1_suppl), 19–31. https://doi.org/10.1111/j.1467-954X.2010.01909.x
Dupré, J. (2012). *Processes of life: Essays in the philosophy of biology*. Oxford University Press.
Dupré, J. (2013). I—John Dupré: Living causes. In *Aristotelian society supplementary volume* (Vol. 87, No. 1, pp. 19–37). Oxford University Press.
Dupré, J. (2015). Pluralism and process in understanding human nature. *Rivista di filosofia neoscolastica, 1/2*, 15–24.
Dupré, J. (2018a). Processes, organisms, kinds, and the inevitability of pluralism. In O. Bueno, R.-L. Chen, & M. B. Fagan (Eds.), *Individuation, process, and scientific practices* (online ed.). Oxford Academic. https://doi.org/10.1093/oso/9780190636814.003.0002
Dupré, J. (2018b). Human nature: A process perspective. In E. Hannon & T. Lewens (Eds.), *Why we disagree about human nature* (online ed.). Oxford Academic. https://doi.org/10.1093/oso/9780198823650.003.0006
Dupré, J. (2024). The disunity of science and the unity of the world. Presidential address, PSA 2022. *Philosophy of Science, 91*, 1043–1057.
Dupré, J. (2025a). *Everyone flows: A process philosophy of human life*. Oxford University Press.
Dupré, J. (2025b). Disorder. *Theoria, 40*(1), 5–16. https://doi.org/10.1387/theoria.27510
Dupré, J. (2025c). Process. *Theoria, 40*(1), 17–28. https://doi.org/10.1387/theoria.27512
Dupré, J., & Guttinger, S. (2016). Viruses as living processes. *Studies in History and Philosophy of Science Part C: Studies in History and Philosophy of Biological and Biomedical Sciences, 59*, 109–116.
Dupré, J., & O'Malley, M. A. (2009). Varieties of living things: Life at the intersection of lineage and metabolism. *Philosophy and Theory in Biology, 1*. https://doi.org/10.3998/ptb.6959004.0001.003
Ereshefsky, M. (Ed.). (1992). *The units of evolution: Essays on the nature of species*. MIT Press.
Ereshefsky, M. (2022). Species. In E. N. Zalta (Ed.), *The Stanford encyclopedia of philosophy* (Summer 2022 ed.). https://plato.stanford.edu/archives/sum2022/entries/species/
Fodor, J. A. (1980). Special sciences, or the disunity of science as a working hypothesis. In *The language and thought series* (pp. 120–133). Harvard University Press.

Ghiselin, M. T. (1974). A radical solution to the species problem. *Systematic Biology, 23*(4), 536–544.
Hull, D. L. (1976). Are species really individuals? *Systematic Zoology, 25*(2), 174–191.
Hull, D. L. (1988). *Science as a process: An evolutionary account of the social and conceptual development of science*. University of Chicago Press.
Keller, E. F. (2000). *The century of the gene*. Harvard University Press.
Kripke, S. A. (1980). *Naming and necessity* (Vol. 217). Blackwell.
Nicholson, D. J., & Dupré, J. (2018). *Everything flows: Towards a processual philosophy of biology* (p. 416). Oxford University Press.
Okasha, S. (2002). Darwinian metaphysics: Species and the question of essentialism. *Synthese, 131*(2), 191–213.
Oppenheim, P., & Putnam, H. (1958). Unity of science as a working hypothesis. *Minnesota Studies in the Philosophy of Science, 2*, 3–36.
Putnam, H. (1975). The meaning of "meaning". *Philosophical Papers, Mind, Language, and Reality, 2*, 215–271.
Simons, P. (2000). Continuants and occurrents. In *Aristotelian society supplementary volume* (Vol. 74, No. 1, pp. 59–75). The Aristotelian Society.
Simons, P. (2018). Processes and precipitates. In D. J. Nicholson & J. Dupré (Eds.), *Everything flows: Towards a processual philosophy of biology*. Oxford University Press.
Smart, J. J. C. (1959). Can biology be an exact science? *Synthese, 11*(4), 359–368.
Stout, R. (2016). The category of occurrent continuants. *Mind, 125*(497), 41–62.
Veigl, S. J., & Currie, A. (Eds.). (2025). *Methods in the philosophy of science: A user's guide*. MIT Press.

Part I
Promiscuity and Disorder

Chapter 2
Unearthing Soil Kinds: How Values and Practices Affect Soil Classifications

Catherine Kendig

Abstract What is useful about considering a metaphysics of soil and what does it tell us about what soil is as well as what we should make of it? Reviewing some of the ways in which the concept of soil health has been defined and how the concept has been operationalized reveals that how soil is conceptualized and classified is inextricably entwined with how soil is managed, who is managing it, and for what purposes it is managed. While a metaphysically neutral account of soil may appear possible, most soil classifications rely on value-laden commitments to what soil is, can or should be. The purposes for which different soil classifications are intended functionally define what are treated as salient soil interactions and which features of soil are appropriate indicators for use in soil assessments. Evaluation of soils relies on different classification systems for different purposes whether they are classifications of soils for geotechnical engineering, for agricultural capability, or for ecosystem services. A pluralist account of soil classification is not just based on the different use made of soil but is grounded in a processual metaphysics. Soil is a collaborative process whose existence depends on the sum of coordinated efforts of plants, microbes, macrobes, non-human animals, and human interventions and whose activities contribute to soil metabolism and other soil capabilities.

2.1 Why Does Classification Matter?

What sort of thing is soil? Is soil a process, an entity, a set of relationships, or a system? I investigate this metaphysical question by focusing on *soil health*, a concept widely used in the agricultural sciences. What makes soil healthy requires understanding several intertwined causal interactions, including microbial and macrobial plant-soil relationships, and the biological and chemical interactions involved in soil metabolism and nitrogen cycling. For this knowledge to be useful, it needs to be

C. Kendig (✉)
Department of Philosophy & Department of Ecology, Evolution and Behavior, Michigan State University, East Lansing, MI, USA
e-mail: kendig@msu.edu

S. Leonelli et al. (eds.), *Life Beyond Human Boundaries: John Dupré's Philosophy and the Interdisciplinary Study of the Life Sciences*,
https://doi.org/10.1007/978-3-032-26977-5_2

operationalized. The characteristics and constituents salient to soil health need to be identified and measured. But identifying which features are salient relies on both metaphysical commitments and normative considerations. That is, they rely on what soil is thought to be and what soil is thought to be for. A metaphysics of soil does not yet exist in philosophy. But, before John Dupré's empirically-informed reconceptualizations of natural kinds and classification, nor did a fleshed-out metaphysics of microbes, macrobes, or metabolism.

One might reasonably ask, why it is important to know what sort of thing soil is—or for that matter what sort of thing anything is in the world. That is, why is the categorization of stuff important at all? Sorting stuff into categories is important not just for a tidy and useable work shed, but also for allowing us to know things about the stuff we sort as well as the as-yet unsorted stuff. Classifying what we know or what we think we know by means of categories and kinds is necessary for a wide range of scientific practices from hypothesizing and testing to the innovative development of new technologies. Why? In general terms, knowing that something is an entity, process or system provides knowledge of it by virtue of its inclusion in that particular category. More specifically, knowing that something is a member of a particular category of living entities rather than another, tells us something about its mode of living and other features of its existence. Natural kinds and their nature as such have, of course, a very long history in philosophical thinking.

Analytic philosophers conceived of wonderful and inventive thought experiments by which to analyze what might be essential features of natural kindhood and the fundamental nature or naturalness of natural kinds. But what, if anything, can these tell us about the accuracy of our scientific classifications and whether they pick out natural kinds? Are the examples long relied upon as quintessential natural kinds (e.g. tigers, gold, water), indeed natural kinds? To test the accuracy of a certain set of categorizations long thought to be foundational in philosophy, Dupré swaps the analytic philosophers' thought experiments for actual biological ones and asks: what happens when we bring empirical understanding of biological taxa to bear on traditionally conceived of natural kinds? In "Natural Kinds and Biological Taxa" (Dupré, 1981, p. 74), and later in *The Disorder of Things* (Dupré, 1993), he provides impetus for rethinking the nature of natural kinds not just for philosophers of biology interested in metaphysics, but for metaphysics *ut totum.* In his discussion of natural kinds, Dupré invites his reader to "consider the lilies":

> Species which are commonly referred to as lilies occur in numerous genera of the lily family (Liliaceae)…I would not want to undertake the task of describing the taxonomic extension of the English term 'lily'. However, it is fairly clearly well short of including the entire family. To include the onions and garlics (genus Allium)…would surely amount to a debasement of the English term. (1981, p. 74)

He later argues that the questionable nature of the natural kindhood of the species of lilies in Liliaceae is not just a problem with language. Even if we are able to invent a grammatical fix that permits our use of the term "lily" and that also rescues its natural kindhood, a much more distasteful problem awaits us.

> It would be a severe culinary misfortune if no distinction were drawn between garlic and onions. But we have seen that this is not a distinction reflected in scientific taxonomy. (1981, p. 80)

For the gourmet, knowing something is an onion rather than garlic is salient for the purposes for which this knowledge is used. To be sure, that onions are lilies according to scientific taxonomy is a classification of little use to the gourmet. The classification problem that Dupré reveals is that, *pace* Putnam,[1] there are no privileged sameness relationships that exists to distinguish species that do not rely on the interests of those seeking to classify. Which distinctions are considered salient restricts the set of properties that are used to make those distinctions. The salience of these distinctions and the properties employed to make these assessments are dependent on both interest and perspective—not just of the gourmet, but also of biologists.

At a time when doing metaphysics was often synonymous with doing philosophy of language and when philosophy of biology was all about selectionism, Dupré showed not only why philosophers of biology needed to do more metaphysics but why metaphysicians needed to consider actual rather than merely fictional biological examples. Serious consideration of how classification works and the consideration of the taxonomic status of lilies in discussions of natural kinds wasn't just a problem for philosophers of biology interested in the species problem or bemused garlic-bothering foodies. It was a necessary line of investigation for both philosophy of biology and traditional metaphysics.

Dupré showed why these commitments and values not only provide justification for knowledge, but also ground biological metaphysics. Dupré didn't just tell us about it—he showed us in countless case studies in everything from ecology to cultural anthropology and from socioeconomics to genomics. In doing so, he inspired so many of us budding philosophers of biology at the ESRC Centre for Genomics in Society (Egenis) to start investigating philosophical areas of research in biology—even more broadly construed than had been before—which had, until then, received little to no attention in philosophy of science.

2.2 Natural Kinds, Kinders, and Kinding Activities

Being part of the Egenis research community and earning my PhD in Philosophy from the University of Exeter in 2009, meant most of what I was thinking about at any moment was the historical and philosophical bases for classification broadly construed and the weird and wonderful ways genomes, organisms, species, and symbionts, as well as their analogous and homologous relationships and reticulated lineages misbehaved or otherwise made trouble for the categorizations to which

[1] Dupré's botanical and culinary examples challenge Putnam's assumption that ordinary language corresponds to any recognized biological taxon and that there is one unambiguously correct taxonomic theory.

they were often conceived to belong. Dupré and the highly interdisciplinary community of Egenis faculty and graduate students helped me consider the metaphysical, ontological, and epistemological significance of genomic, proteomic to metabolomic research, how to reconceive species concepts and reimagine the species problems in light of misbehaving species, consider reticulated evolutionary lineages, eco-evo-devo, synthetic and systems biology, and have continued to inspire me to pursue research in many lesser studied areas in philosophy of science.

What unifies some of the most important philosophical thinking about classification of whatever it is that one is considering, is that if we want to take metaphysical considerations seriously—we need to question presumed ontologies. Classifications often smuggle in metaphysical assumptions which make devising hypotheses possible, testing seem to follow from the devised hypotheses, and causal explanations make sense. To reveal these requires challenging the long-held assumptions and asking: what is it that makes natural kinds natural and why does it even matter?; can realists be realists without assuming the mind-independence claim of traditional approaches?; or what is the evidence for thinking that there is one univocal monist theoretical concept of species instead of a plurality of non-concordant concepts?

These are metaphysical questions which outstrip the armchair. Whether a pluralist metaphysics fits a disunified disorderly world or whether a causal explanation is fit for purpose is something that requires empirical examination. If we want to understand how classification works, we need to investigate how and why folks are classifying as well as what they are classifying and how they are doing it successfully. This practice-shaped metaphysics includes investigating classifying, reclassifying and unclassifying activities as well as conceptualizing and reconceptualizing methods of analysis. It is a metaphysical examination that is robustly interdisciplinary and so requires active engagement with specialists. Why? Because, harkening back to Dupré's invitation to "consider the lilies" means considering the properties of being a lily or an onion that are not those recognized in biology but are recognized by growers of lilies and onions, appreciators of lilies, and eaters of onions.

Building on Dupré's work on natural kinds, classification, and the metaphysics of biology, I began to investigate agronomists', gardeners', cooks', and farmers' activities of researching, growing, tending, and interacting with not just lilies but other agricultural species and varieties and the causally relevant practices and environments required in the making of them. I turned the focus of natural kinds discussions from natural kinds to the *kinders* of kinds, and the activities of *kinding*, and reconceived of kinds as that which are produced as the result of kinding activities (Kendig, 2016a, 2016b, 2020, 2024; Kendig & Bartley, 2019; Kendig & Grey, 2021). In "Activities of kinding in scientific practice", I first describe what I mean by "kinding activities":

> "*Kinding* activities" refers to the different sorts of practices individuals and communities use in making, discovering, delimiting, reconstructing, and sharing the kinds that are of interest to them. (2016a, pp. 1–6)

Kinding activities are those practices used by a community to identify and reidentify those entities or processes of meaning and value to it. Kinding activities anchor the kindness of kinds in a particular community. As such, they connect the ontological category of kinds with how they are instantiated for different communities holding different ontological commitments and different value-laden epistemic aims. These kinds are real, but their reality is not grounded in some inaccessible noumenal world out of our reach, but in the promiscuous methods by which we access them. Refocusing discussion of natural kinds on kinders and kinding activities is intended to foreground the role of the knower and classifier in biological metaphysics. In the following, I bring the focus on the knower and classifier to examine categorizing practices in a much less discussed area of philosophical study, philosophy of agriculture, and in particular philosophy of soil classification.

2.3 Normative Categorization in the Agricultural Sciences

Agricultural categories provide normative evaluations of lands capable to be farmed, the food standards which must be met in order for a product to be sold as such, and what are the appropriate farming practices that should be followed by growers. Agricultural lands are classified in terms of capability units and farms are evaluated in terms of their yield potential. Agricultural products are produced according to prescribed standards of care and graded prior to becoming marketable goods. Whilst farming practices are judged as being environmentally sustainable or those leading to ecological degradation.

One set of categories and standards widely discussed in the agricultural sciences are those concerning soils and their management. This is perhaps unsurprising. Soils give us 95% of the food we eat, but they also contain 1,460 billion tonnes of organic carbon—the largest terrestrial carbon reservoir and twice that of atmospheric carbon (Vargas, 2022). One gram of soil contains from 10,000 to 52,000 different microbial bacterial species (or operational taxonomic units roughly translating to the species level) (Roesch et al., 2007, pp. 283–285). Soil management relies on several cross-cutting soil classifications. Soil is classified according to its physical properties and chemical and biological constitution by soil surveys, soil types, and soil capability units.

Categorizations of soil in agriculture focus on what soil is for and what are the chemical and biological features it must have to do that successfully. Top of the list of these normative categorizations of soil is *soil health*. The concept of soil health has been widely used in agriculture and conservation for the past twenty-five years (Doran & Zeiss, 2000). While the term "soil health" is relatively new, the conceptualization of the well-being of the soil ecosystem has a long history with soil being described in terms of "fertility", "goodness", and "vitality" (Brevik & Hartemink, 2010; Carson, 1962; Lovelock, 2000). To understand the concept of soil health and what it entails requires unearthing some of the metaphysical, epistemological and methodological commitments underpinning it. For instance, what is soil, what are the measurable indicators of soil health, and what can be changed to make soil healthy.

2.3.1 What Is "Soil Health"

Over the last century, discussions of soil fertility and quality have brought together biological and chemical knowledge, knowledge of plant physiology, and anthropological and sociological information gleaned from the activities and motivations of generations of farmers' management decisions each year. These have assessed several microbiological activities of the soil community including soil metabolism, and the macro and micro above and below ground soil, plant, and animal interactions. Soil fertility has been defined as a quality or attribute of soil, a process, and a goal that directs farm management activities and shapes agricultural research questions.

Despite being in widespread use in much of the recent research in soil biology and biochemistry, farming, agronomy, and conservation research, soil health terminology, the conceptualization of soil health, and its operationalization have been a subject of frequent debate. Some of the debate centers on whether the term is a metaphor, a concept relating to actual soil processes or capacities, or a concept that can shape knowledge of soil goals such as sustainability. For instance, Janzen and colleagues (Janzen et al., 2021) argue that soil health is merely a metaphor and a not very good one at that. While the term is especially popular in soil and environmental sciences, they suggest we should be wary of its use: "Despite its ubiquity, the concept remains elusively ambiguous, largely because 'health' here is a metaphor, not a literal scientific construct" (Janzen et al., 2021, p. 1). Their solution is to redefine soil health as: "the vitality of a soil in sustaining the socioecological functions of its enfolding land" (Janzen et al., 2021, p. 1). Doing so appears to be a shift in the subject of discussion—from soil to land. A move that some might suggest does not remove the ambiguity of what soil health is, but simply increases the scope of what it could mean.[2] Janzen and colleagues justify limited use of soil health as a metaphor. They argue that despite being a metaphor, the term can be used to motivate discussion of the interrelationships with soil and ecosystem functions and stimulate transdisciplinary views of soil than include micro- and macrobiotic organisms and their biotic and abiotic interactions. Others, such as Lehmann and colleagues, argue that the concept is not metaphorical but a concept that accurately represents a host of soil-relationships that can potentially apply to different scales. For Lehmann and colleagues, soil health is not a metaphor but a versatile platform concept that serves to represent current and potential properties of soil complexes; "soil health is an overarching principle which contributes knowledge, [and] as a property to measure environmental and societal functions" (Lehmann et al., 2020, p. 552). While consensus on what the soil health concept means and to what it refers remains a topic of active debate, its assessment by means of soil tests continues.

Different soil tests assess the physical, chemical and/or biological indicators of soil health and continue to be the main route by which soil is managed by farmers and understood by researchers. While categorization of these indicators as either physical, chemical or biological is common, the properties and capabilities of soil that are

[2] See, for instance, pluralist alternatives offered by Lehmann et al. (2020) and Powlson (2020), which are discussed later in this chapter.

sought to increase yield are known by both farmers and researchers to be the result of multiple interacting multiscale processes. The tripartite classification of chemical, physical or biological is not a categorization of discrete sources of soil health but of the kinds of soil constituents whose measurement is possible and whose manipulation makes an agriculturally or environmentally significant difference to the soil system of which it is a part. For instance, measuring the amount of nitrogen in the soil system is a chemical bellwether for soil health or degradation. The amount of nitrogen in soil is the result of several potential anthropogenic causes: the chemical nitrogen amendments added by farmers and the biologically produced nitrogen resulting from the microbial activities of nitrogen fixing bacteria in leguminous crops[3] that provide nitrogen to other non-nitrogen fixing plants.

2.3.2 Soil Health as Land Health

Defining capacities of soil health, soil fertility or soil goodness in terms of a more holistic notion of land health is of course not new. Aldo Leopold's notion of *land health* as defined in his "Conservation: In whole or in part" is used as a collective concept of "land-as-a whole" (Leopold, 1944, p. 310). In "The Land Ethic", *land health* was also interpreted as a capacity possessed by it—and not as a mere metaphor. Land health is understood as the potential for stability of a system or its sustainability, "health is the capacity of the land for self-renewal" (Leopold, 1949, p. 221). *Conservation* is defined in terms of "our effort to understand and preserve this capacity" (Leopold, 1949, p. 221). But land health, like its conceptual descendent soil health, was also a concept widely used but widely debated. Leopold concedes there is extensive disagreement across different disciplines, such as forestry, agronomy, wildlife studies, agriculture, ecology, and farming in what "the land" includes, what it does not (Leopold, 1949, pp. 221–222).[4] In his later work, Leopold explicitly extends his notion of land health to include the integral organizational relationships of soil, water, animals and plants. For Leopold, land health was "a fountain of energy flowing through a circuit of soils, plants, and animals. Food chains are the living channels which conduct energy upward; death and decay [back] to the soil" (Leopold, 1949, p. 216). Land health was not something that could be understood by means of a reduction of land or soil to its microphysical description but instead to the activities of a system. Leopold's own redefinitions and reimagined boundaries of what land health is, to what does it refer, and how it obtains appears to prefigure some of the conceptualization(s) of soil health now being debated.

[3] Leguminous crops include alfalfa, white clover, red clover, soybeans, broad beans, chickpeas, and peanuts to name a few. These can be planted either as cash crops or as cover crops.

[4] See Millstein (2024) and O'Neill (2001) for further discussion of Leopold's own shifting concept of land health, especially in his writings from 1934 to 1949.

2.4 Soil Classification: How Is Soil Classified and for What Purposes?

In the last section, I argued that the chemical, physical and biological properties of soil should not be considered to be discrete sources of soil health but that they instead refer to different kinds of soil constituents whose measurement makes a difference when treated as a potentially manipulatable part of the soil system. I want to return to this point and unpack what this means for the concept of soil health. We might start by asking, what is the concept of soil health meant for? A quick answer to this that I mentioned in passing in the previous paragraphs, is that the assessment of soil health already assumes what sort of thing soil is for in order to assess the suitability of soil health indicators. The choice of soil properties deemed suitable as soil health indicators includes the identification of which properties are those that can be intervened on. Accurate and informative identification of which properties should be those chosen to serve as reliable measures of soil health that can inform better soil management must be those which are capable of change given certain intervention strategies. A longer answer will require bringing two of the topics discussed so far into conversation with each other to first answer the question: what does soil management have to do with soil classification?

First of all, how is soil classified? Soil tends to be classified in terms of what is being done with it and as such different classification systems apply to soil used in engineering and maintenance of roads and in agriculture. In both, soil can be classified in terms of its geological origin, physical structure, and grain size. There are also unified soil classification systems and preliminary classifications of soil types, ones that rely on local knowledge or colloquial names of regional soils, and ones assessing soil capability for agricultural activities. For instance, often a unified soil classification system is sought for use in engineering and geology to describe the texture of soil as well as the grain size of its particulate matter in terms of what percentage of its weight passes through graded sieves.[5] The American Society for Testing and Materials (ASTM International) developed the Standard Practice for Classification of Soils for Engineering Purposes called the Unified Soil Classification System. This unified system classifies soils from "any geographic location into categories representing the results of prescribed laboratory tests to determine the particle-size characteristics, the liquid limit, and the plasticity index" (ASTM, 2020, p. 3) [see Fig. 2.1]. The ASTM USCS is designed to classify soils according to the geotechnical engineering behavior of soils useful in lab or field experiments.

Contrast this with the 1961 United States Department of Agriculture's (USDA) "Land-Capability Classification" devised by A. A. Klingebiel and P. H. Montgomery. This classification system provided groupings of soil into categories based on their use and management into soil mapping units, capability units, capability subclasses, and classes focusing on four general kinds of limitations: erosion, wetness, rooting zone limitations and climate (Klingebiel & Montgomery, 1961, p. 3) [see Fig. 2.2].

---

[5] This was based on the Airfield Classification System developed by Casagrande in the 1940s and widely adopted by 1952 as the Unified Soil Classification System (ASTM, 2020).

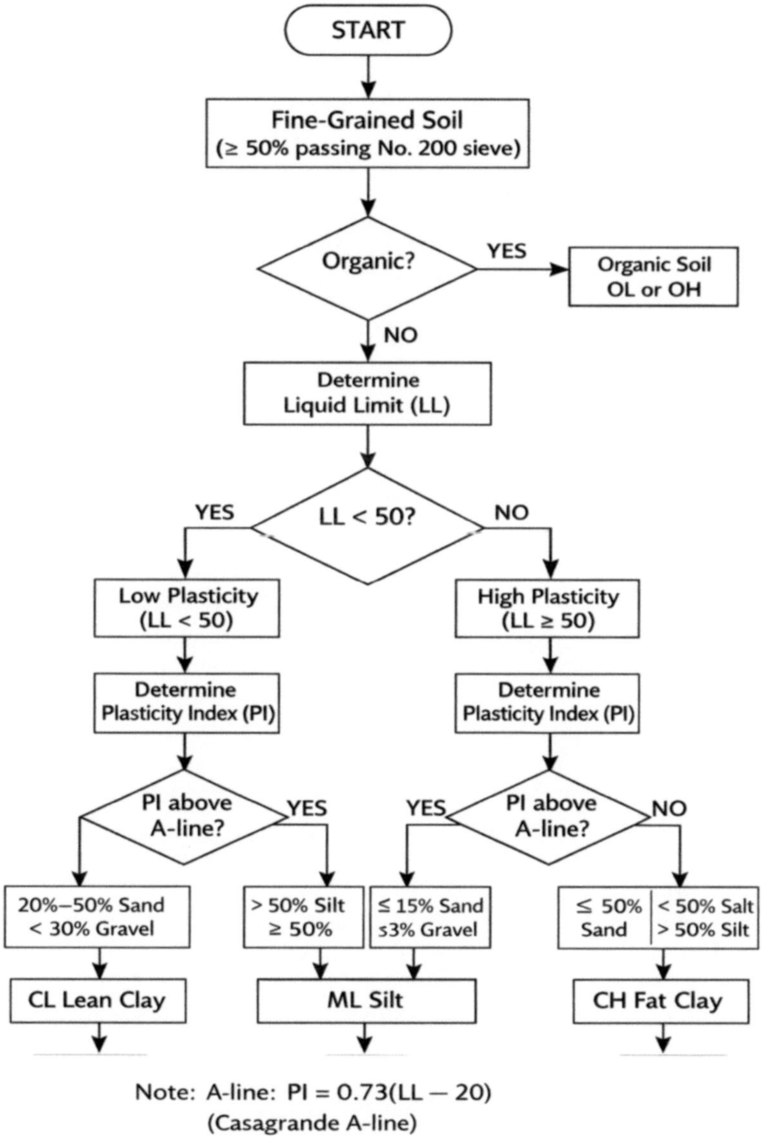

Fig. 2.1 A simplified flow chart showing the standard practice for classifying fine-grained soil for engineering purposes (≥50% passing through no. 200 sieve). Drawn by the author after the ASTM Unified Soil Classification System

Soil-mapping unit	Capability unit	Capability subclass	Capability class
A soil mapping unit is a portion of the landscape that has similar characteristics and qualities and whose limits are fixed by precise definitions. Within the cartographic limitations and considering the purpose for which the map is made, the soil mapping unit is the unit about which the greatest number of precise statements and predictions can be made. The soil mapping units provide the most detailed soils information. The basic mapping units are the basis for all interpretive groupings of soils. They furnish the information needed for developing capability units, forest site groupings, crop suitability groupings, range site groupings, engineering groupings, and other interpretive groupings. The most specific management practices and estimated yields are related to the individual mapping unit.	A capability unit is a grouping of one or more individual soil mapping units having similar potentials and continuing limitations or hazards. The soils in a capability unit are sufficiently uniform to (a) produce similar kinds of cultivated crops and pasture plants with similar management practices, (b) require similar conservation treatment and management under the same kind and condition of vegetative cover, (c) have comparable potential productivity. The capability unit condenses and simplifies soils information for planning individual tracts of land, field by field. Capability units with the class and subclass furnish information about the degree of limitation, kind of conservation problems and the management practices needed.	Subclasses are groups of capability units which have the same major conservation problem, such as— e—Erosion and runoff. w—Excess water. s—Root-zone limitations. c—Climatic limitations. The capability subclass provides information as to the kind of conservation problem or limitations involved. The class and subclass together provide the map user information about both the degree of limitation and kind of problem involved for broad program planning, conservation need studies, and similar purposes.	Capability classes are groups of capability subclasses or capability units that have the same relative degree of hazard or limitation. The risks of soil damage or limitation in use become progressively greater from class I to class VIII. The capability classes are useful as a means of introducing the map user to the more detailed information on the soil map. The classes show the location, amount, and general suitability of the soils for agricultural use. Only information concerning general agricultural limitations in soil use are obtained at the capability class level.

Fig. 2.2 Relationship of soil-mapping unit to land capability classification (Klingebiel & Montgomery, 1961, p. 2)

According to the 1961 Land-Capability Classification, Class I soils are described as "soils that have few limitations that restrict their use… are nearly level, well drained, have deep rooting zones, and are easily maintained in good tilth" (Klingebiel & Montgomery, 1961, p. 6). Whereas soils categorized in Class III were described as those that are significantly prone to erosion and have "severe limitations" restricting both what could be planted and when (Klingebiel & Montgomery, 1961, p. 8).[6] The classification of soils for engineering purposes and the capability classification of soil for agricultural purposes provide just two of many normative soil classifications and rely implicitly or explicitly on different soil particle properties such as structural or chemical properties, soil aggregate or system-level properties (such as erosion or drainage). While different, they both have been devised for (different kinds of) soil management and soil investigations.

Soil health categorizations present a third sort of normative classifications in addition to the two just mentioned (i.e. the ASTM Unified Soil Classification System and the USDA Land-Capability Classification) that have been devised for the purpose of guiding soil management decisions. In addition to physical properties and agricultural capability, soil health categorizations may rely on organic properties of soil and soil constituents such as decayed plant matter, organisms and their residue, and the presence of worms. Assessments of soil health may also include the collaborative relationships within the soil system that facilitate communication between plants, soil, and the microbial and macrobial communities by means of soil-root exudate interactions.

[6] For a more in-depth discussion of the use and revision of Klingebiel and Montgomery's Land-Capability Classification and other interacting social ontologies of soil management, see Kendig (2024).

Determining what sorts of knowledge can be gained through these different classifications relies on variable commitments to which features and interactions are identified as salient for which purposes. To understand what soil is as well as what we can conclude from soil science experiments, what is included and what is not included *as being soil thus classified* matters.

2.5 Soil as a Living System Capable of Metabolizing

In discussions of the concept of soil health, the soil or land to which the health refers is often assumed to be a system capable of change over time. This is captured in the definition provided by the Natural Resources Conservation Service of the USDA: "Soil health is the continued capacity of soil to function as a vital living ecosystem that sustains plants, animals, and humans" (Natural Resources Conservation Service/ USDA, 2024). Soil systems are not inert but made up of dynamic trophic interactions between different microbial and microbial communities. Soil microbes eat, respire, release nutrients derived from breaking down organic matter, reproduce, and die. As such, we can safely add soil to the list of systems whose description Dupré and Nicholson would likely say is best characterized and understood as a process rather than as an entity or other metaphysical category of being:

> many processes have boundaries that are fuzzy or indeterminate…processes are individuated not so much by where they are as by what they do. A series of activities constitute and individual process when they are causally interconnected or when they come together in a coordinated fashion to bring about a particular end as integrated systems. (Dupré & Nicholson, 2018, p. 13)

The conception of soil health provided by the NRCS, explicitly refers to soil health as being the result of a "continuing capacity" (a description of a system-wide property). The description implies that soil can at least sometimes be understood as a living system—one capable of being assessed as healthy or sick. Soil, conceived of in this way appears to provide at least some motivation to think it also can be categorized as a continuant process.

To understand how soil can be conceived of as a living system, we need only turn to investigate which are features and interactions which make it a living system. One set of interactions which Dupré and O'Malley argue is crucial to most living systems is metabolism. To consider metabolism as a litmus test for life need not rely on the assumption that metabolism needs to be the contained process of a specific organism.[7] As Dupré and O'Malley argue, life is best understood with a collaborative interpretation on it and many of life processes, such as metabolism, is often

[7] Even when thinking about an individual organism (e.g. a cow, peanut plant, cat or a human), metabolism remains a community effort. Metabolism is the result of the coordinated activities of exosymbiotic and endosymbiotic microbes which extract nutrients from our food, fight infections, and survive in the way they each do together (see Dupré & O'Malley, 2009, pp. 13–14).

a collaborative effort—the result of collaborative exchanges and community interactions. Instead of a set of competing candidate conceptualizations of life, Dupré and O'Malley contend life is best investigated and understood through the lens of cooperation where the intertwining of life cycles, lineages, and metabolic processes is revealed as collectively contributing to living systems;

> the tension between replicating lineages as one criterion of life, and metabolic sustainability as the other, can be reconciled by taking a more interactive view of metabolic processes and by reconceiving cooperation and competition within a broader framework of collaboration—life, according to our analysis—occurs at the intersection. (Dupré & O'Malley, 2009, p. 14)

Soil has long been described as a metabolizing system but also one where the sorts of causal interactions remain contested. While summarizing many of the advances in soil biochemistry in the early 1900s, J. Wrightson and J.C. Newsham pondered,

> The wonderful facts recently established with regard to antagonistic bacteria, protozoa, phagocysts, amoeba, algae, moulds etc., all indicate the presence of a subterraneous flora and fauna which produce all kinds of intermediate compounds within the soil and affect its 'complex particle.' Viewed from the new point of view, who can say what may be the cause of barrenness[8] and unproductiveness in certain cases? (Wrightson & Newsham, 1919, p. 49)

In his Leeuwenhoek Lecture entitled "Soil metabolism" delivered to the Royal Society in 1954, Quastel describes his experiments and observations of some of these intermediate compounds, their causal sources, explaining how they contribute to the potential productivity of soil and its continued capacity to sustain life. Quastel emphasizes that soil systems are not just chemical substrates for plants but are instead biologically constituted in such a way that they are capable of metabolizing. Outlining several metabolic cycles within soil systems, including the process of nitrification, the manganese cycle and sulphur metabolism among others Quastel, describes their interweaving causal impacts on the soil community. For instance, in describing sulphur metabolism, he explains, there are:

> varied species of organisms that arise and are capable of attacking both the initial substrate and the products derived from it. These dependent organisms develop almost simultaneously…forming a cycle where the sulfur acts in a catalytic role because by its varied transformations it secures the growth of the groups of organisms whose energies for development are obtained from oxidations and reductions of specific sulphur-containing substances. (Quastel, 1954, p. 168)

How does soil metabolism affect soil health? Microorganismal activities and their collective deposits of biologically active products give food to other microbes and plants allowing them to grow (cf. Lynch & Audus, 1976, p. 68). These are not all the same activities or products. For instance, some microbes degrade and eat chitin (a carbon and nitrogen source in the soil), whilst others that are unable to degrade chitin, eat other nutrients produced by chitin-degrading and other microbes within the community (McClure et al., 2022). Soil metabolism is also not the same process in different soils. It changes in both the rate as well as the pathways utilized. When

[8] Soil fertility or its barrenness was not a new topic in agriculture, it had been a concern as old as agriculture itself, its barrenness chronicled in every agronomic culture.

adding labelled glucose to different soils, Dijkstra and colleagues (2022, pp. 390–391) found that the microbial communities in freshwater marsh soils, high elevation alpine conifer forest soils and desert grassland soils use different biochemical pathways by which to process the glucose.[9]

2.6 What Kind of System Is a Soil System?

While there is some general agreement that soil health reflects the interacting metabolic cycles of diverse microbial communities. This agreement does not translate to a consensus over how to measure these interactions. Which characteristics of soil are associated with soil health, how these are best measured, what sort of knowledge is gained by means of soil health tests and which interventions would provide—given the information gleaned from soil health tests—the best means by which to increase soil health are all questions whose answers remain very much the topics of much debate. What does seem to be a developing trend in thinking is a growing pessimism that soil health is something that "cannot be directly measured" (Maaz et al., 2023, pp. 4–5). Reliance on highly contextualized measurements of soil constituents and material features of soil such as soil texture, pH, soil organic matter, microbial diversity, hot water extractable carbon, 24 h CO_2 burst, nitrogen mineralization, water holding capacity, water stable mega aggregates, total organic carbon, ß-glucosidase, and ß-glucosaminidase are treated as biological, chemical and physical indicators that are relevant to specific ecosystem services insofar as they play certain functional roles in soil (Maaz et al., 2023, pp. 4–8; Powlson et al., 2015). These soil health indicators are treated as measurable proxies for soil health with no ontological commitment to there being an entity to which soil health is attributed beyond the proxy measures. What is provided in these is not a measure of the health of a soil system but instead a set of measurements of those features of soil that are conceived of as functionally salient in generating certain ecosystem services judged to be valuable. This ontologically agnostic, normatively defined functional approach is not something unique to soil. What is understood about various features of agricultural landscapes is often the result of what is considered to be valuable for a particular purpose given a particular endeavor. The problem with this ontologically agnostic approach is that it unquestioningly prioritizes the human point of view in valuing ecosystem services. Arguably identifying functionally salient soil features as indicators of soil health necessitates determining their functional salience from *a* particular point of view, however this view needn't be always assumed to be a human one. For instance, some constituents of soil—for instance, trace amounts of Boron—might turn out to be toxic for some species of microorganisms, while for

[9] Dijkstra and colleagues found that microbial communities in marsh soils used Embden-Meyerhof-Parnas glycolysis, the conifer forest soil relied upon Entner-Doudoroff and pentose phosphate pathways, and the grassland soil relied on Entner-Doudoroff pathway (Dijkstra et al., 2022, p. 391 fig 4).

plants, it is an essential element for nutrition (Brenchley, 1947). Whilst a farmer intending on growing maize may regard her sandy nutrient poor soil as unhealthy, the lavender in the neighboring field with the same sandy soil may be flourishing. From the point of view of the lavender, the soil is healthy as it provides the soil that lavender needs to grow.

In contrast to this functional approach, others seem more ontologically committed to the view that soil health does indeed refer to something—it refers to the soil system as a whole. On this view, soil health is an emergent set of properties along with life, function, complexity, and resilience of a soil system (Harris et al., 2022, pp. 2–3). A systems approach not only commits itself to the ontology of systems, it is also committed to there being features and characteristics of systems *qua* systems. That is, a systems approach can be conceived of as one that does not merely treat features of systems as those which are of functional interest to humans, plants or microbes, but as features of systems at a particular scale of analysis. As such, the concept is not used as a metaphor that borrows from our understanding of human health. Instead, it is an explicitly non-human notion of health. While I am sympathetic to the position Harris and colleagues outline, I think their reference to emergent properties is much too quick.

2.7 Identifying Actionable Indicators of Soil Health

In order to better understand how soil health can be emergent from the activities of a soil system, one possible route could be to adopt Dupré's conception of living causes within his processual approach to development systems where "development emerges from a sequence of interactions between biology and environment" and "causation involves the intervention in a process" (Dupré, 2013, p. 32). Doing so would also help connect the soil health concept to what soil health information is for and what to do with soil to make it healthier. Answers to these questions do not just rely on how conceptualizations of soil health are operationalized. Which characteristics are identified as being indicators of soil health varies depending on who is seeking to assess soil and what they are doing with it. Soil health conceptualizations, the interpretations of soil health data, and in particular soil biological health indicators, appear to differ across different communities of farmers and researchers (e.g. cotton farmers, potato farmers, agronomists, horticulturalists, and agbiotech developers) (Ingram et al., 2010; Singh et al., 2023; Wade et al., 2021). The lack of standardization in soil biological health indicators means that what can be concluded from these indicators and the tests designed to assess them is less than clear especially when making decisions about a particular soil management strategy. Knowing the classification of every microorganism in the soil may be informative to a soil microbiologist interested in how methanogenic archaea and ammonia-oxidizing archaea contribute to carbon cycling (see for instance Jung et al., 2020). But this may not be informative to a farmer interested in making a decision between whether she should apply phosphorus, nitrogen, or potassium (or other soil amendments) to her field, or whether planting red clover or rye as a cover crop will be most beneficial for her soil. A soil test that provides information about a general functional feature, such as the microbial biodiversity of the field might. But the usefulness of this general functional feature relies on significant interpretation, such as translation of metagenomic sequencing data, in order to figure out what impact it has on which metabolic cycle that may be of use in farmers' choices between different chemical or biological soil amendments (Graham et al., 2016). Farmers' definitions of soil health and their preference for certain soil indicators over others differs from those of soil researchers for an obvious reason: "many soil biological health indicators have been designed for research purposes and are currently difficult to interpret as actionable outcomes for farmers" (Singh et al., 2023, p. 3). This returns us once again to the question raised earlier in this section: what is the information from soil health measurements for, and who is it for? If soil health measures are to be informative to farmers, they must provide a set of indicators that are more operationally useful to them when they are making soil management choices (Singh et al., 2023, p. 3). In order for soil health tests to provide meaningful data that can be used in soil management decisions, farmers need to be able to map the information provided by soil health indicators on to what they already know about the soil on their farms.

Soil health measurements—viewed in this way as measures which are informative for farmers' soil management decisions—must be value-laden and purpose-directed to be useful. If the goal is to acquire information which will help guide better soil management, then those indicators that reveal deficits in the soil that can be corrected by adopting different soil management strategies will be those that are most useful[10] As such, soil health is a concept that is ontologically, epistemologically and normatively entangled with the soil management.[11] Soil health implies a state of soil that could be better and an implicit goal to identify those practices to improve it. Practices that have been identified as those which can lead to improved soil health include data-driven soil-responsive methods tailored to care for crops and soil such as specific crop rotations,[12] planting of the right cover crops at the right time given the needs of the soil, and precision agricultural management of pest control and nutrient application.[13] Soil health measurements are regarded as being informative for certain

[10] For instance, the farmer may correct the deficits in the soil by minimizing soil disturbance by adopting a no-till strategy, she might choose to maximize the protection and coverage of soil by planting cover crops to prevent erosion or soil run off, she may maximize planting decisions that increase soil biodiversity (e.g., planting perennial prairie strips or widening hedgerows).

[11] Elsewhere I define soil management as the intentional planned interactions with soil guided by underlying commitments to what soil is, what it is thought to be for, and what should be done with it (Kendig, 2024).

[12] Crop rotations refer to a planned sequence of planting crops in the same soil over a determined timeframe. Crop rotations that include legumes like peanuts or red clover provide nitrogen-poor soil with additional biological nitrogen. If the soil has an excess of nitrogen, planting crops that scavenge this extra nitrogen like rye or sorghum can help to prevent the leaching of nitrogen which causes nutrient loss and environmental contamination.

[13] Much like pharmacogenomics or evidence-based approaches to human medicine, tailored agricultural interventions rely on DNA sequencing and molecular profiling to guide specific applications of soil amendments replacing uniform one-size-fits-all approaches.

purposes by those using them. The useability of them as reliable indicators of soil health for these purposes relies on users' ontological commitments to what sort of thing soil is and epistemological, methodological, and normative commitments that inform their management choices regarding how soil should be cared for to make it better.

2.8 Soil Classifications Rely on Value-Laden Commitments

What is useful about considering a metaphysics of soil and what does it tell us about what soil is as well as what we should make of it? Reviewing some of the ways in which the concept of soil health has been defined and how the concept has been operationalized reveals that how soil is conceptualized and classified is inextricably entwined with how soil is managed, who is managing it, and for what purposes it is managed. While a metaphysically neutral account of soil may appear possible, most soil classifications rely on value-laden commitments to what soil is, can or should be. The purposes for which different soil classifications are intended functionally define what are treated as salient soil interactions and which features of soil are appropriate indicators for use in soil assessments. Evaluation of the value of soils relies on the different human points of view relying on classification systems for different purposes whether they are classifications of soils for geotechnical engineering, for agricultural capability, or for ecosystem services. A pluralist account of soil classification is not just based on the different use made of soil but is grounded in a processual metaphysics. Soil is a collaborative process whose existence depends on the sum of coordinated[14] efforts of plants, microbes, macrobes, non-human animals, and human interventions and whose activities contribute to soil metabolism and other soil capabilities.

Acknowledgements Many thanks to Sabina Leonelli, Celso Neto and Stephan Guttinger for inviting me to the Understanding Life in a Changing Planet Conference at the University of Exeter and for putting this volume together. Special thanks to them and to everyone in attendance for their enthusiasm and feedback on an early version of this paper. Thanks also to Aja Watkins and the students in her graduate seminar for their helpful comments on a later written draft.

I want to take the opportunity to express how grateful I am for the vibrant extended lineage of Egenis scholars, both for their friendship and for their invaluable support throughout my PhD and beyond. But of course, my immeasurable gratitude goes to John Dupré, without whom philosophy of biology would not have become the welcoming and inclusive interdisciplinary field that it is, nor I the philosopher of biology I am.

This research is based upon work supported by the National Science Foundation under Grant No. 2240749, "Epistemic and Ethical Functions of Categories in the Agricultural Sciences" (Principal Investigator: Kendig).

[14] "Coordination" here does not mean that the interactivity between plants, microbes, macrobes, non-human animals, and humans is intentional or that the coordination results in either a good or bad state of affairs, but simply that their activities are causally intertwined.

Works Cited

American Society for Testing and Materials (ASTM). (2020). *ASTM D2487–17: Standard practice for classification of soils for engineering purposes (Unified soil classification system).* Book of Standards, Vol. 04.08. Developed by Subcommittee D18.07. https://doi.org/10.1520/D2487-17

Brenchley, W. (1947). The essential nature of certain minor elements for plant nutrition. *Botanical Review, 13*(4), 169–193.

Brevik, E. C., & Hartemink, A. E. (2010). Early soil knowledge and the birth and development of soil science. *CATENA, 83*, 23–33.

Carson, R. (1962). *Silent Spring*. Houghton Mifflin.

Dijkstra, P., Martinez, A., Thomas, S., Seymour, C., Wu, W., Dippold, M., Megonigal, P., Schwartz, E., & Hungate, B. (2022). On maintenance and metabolisms in soil microbial communities. *Plant and Soil, 476*(1–2), 385–396.

Doran, J. W., & Zeiss, M. R. (2000). Soil health and sustainability: Managing the biotic component of soil quality. *Applied Soil Ecology, 15*, 3–11.

Dupré, J. (1981). Natural kinds and biological taxa. *Philosophical Review, 90*(1), 66–90.

Dupré, J. (1993). *The disorder of things*. Harvard University Press.

Dupré, J. (2013). Living causes. *Proceedings of the Aristotelian Society Supplementary, 87*(1), 19–37.

Dupré, J., & Nicholson, D. (2018). A manifesto for a processual philosophy of biology. In D. Nicholson & J. Dupré (Eds.), *Everything flows: Towards a processual philosophy of biology.* Oxford University Press.

Dupré, J., & O'Malley, M. (2009). Varieties of living things: Life at the intersection of lineage and metabolism. *Philosophy, Theory, and Practice in Biology, 1*(3). https://doi.org/10.3998/ptb.6959004.0001.003

Graham, E. B., Knelman, J. E., Schindlbacher, A., Siciliano, S., Breulmann, M., Yannarell, A., Beman, J. M., Abell, G., Philippot, L., Prosser, J., Foulquier, A., Yuste, J. C., Glanville, H. C., Jones, D. L., Angel, R., Salminen, J., Newton, R. J., Bürgmann, H., Ingram, L. J., Hamer, U., Siljanen, H. M. P., Peltoniemi, K., Potthast, K., Bañeras, L., Hartmann, M., Banerjee, S., Yu, R.-Q., Nogaro, G., Richter, A, Koranda, M., Castle, S. C., Goberna, M., Song, B., Chatterjee, A., Nunes, O. C., Lopes, A. R., Cao, Y., Kaisermann, A., Hallin, S., Strickland, M. S., Garcia-Pausas, J., Barba, J., Kang, H., Isobe, K., Papaspyrou, S., Pastorelli, R., Lagomarsino, A., Lindström, E. S., Basiliko, N., & Nemergut, D. R. (2016). Microbes as engines of ecosystem function: When does community structure enhance predictions of ecosystem processes? *Frontiers in Microbiology, 7,* 214.

Harris, J., Evans, D., & Mooney, S. (2022). A new theory for soil health. *European Journal of Soil Science, 73*, e13292.

Ingram, J., Fry, P., & Mathieu, A. (2010). Revealing different understandings of soil held by scientists and farmers in the context of soil protection and management. *Land Use Policy, 27*(1), 51–60.

Janzen, H. H., Janzen, D., & Gregorich, E. (2021). The "soil health" metaphor: Illuminating or illusory? *Soil Biology and Biochemistry, 159*, 108167.

Jung, J., Kim, J. S., Taffner, J., Berg, G., & Ryu, C. M. (2020). Archaea, tiny helpers of land plants. *Computational and Structural Biotechnology Journal, 18*, 2494–2500.

Kendig, C. (2016a). Activities of kinding in scientific practice. In C. Kendig (Ed.), *Natural kinds and classification in scientific practice.* Routledge.

Kendig, C. (2016b). Homologizing as kinding. In C. Kendig (Ed.), *Natural kinds and classification in scientific practice* (pp. 106–125). Routledge.

Kendig, C. (2020). Ontology and values anchor indigenous and grey nomenclatures: A case study in lichen naming practices among the Sámi, Sherpa, Scots, and Okanagan. *Studies in History and Philosophy of Science Part C, 84*, 101340. https://doi.org/10.1016/j.shpsc.2020.101340

Kendig, C. (2024). Human-managed soils and soil-managed humans: An interactive account of perspectival realism for soil management. *Journal of Social Ontology, 10*(2), 80–109. https://doi.org/10.25365/jso-2024-7690

Kendig, C., & Bartley, B. (2019). Synthetic kinds: Kind-making in synthetic biology. In J. R. S. Bursten (Ed.), *Perspectives on classification in synthetic sciences: Unnatural kinds* (pp. 78–96). Taylor & Francis.

Kendig, C., & Grey, J. (2021). Can the epistemic value of natural kinds be explained independently of their metaphysics? *British Journal for the Philosophy of Science, 72*(2), 359–376.

Klingebiel, A. A., & Montgomery, P. H. (1961). *Land-capability classification* (Agricultural Handbook No. 210). Soil Conservation Service, U.S. Department of Agriculture.

Lehmann, J., Bossio, D., Kögel-Knabner, I., & Rillig, M. (2020). The concept and future prospects of soil health. *Nature Reviews Earth & Environment, 1*(10), 544–553.

Leopold, A. (1944). Conservation: In whole or in part? In S. L. Flader & J. B. Callicott (Eds.) (1991), *The river of the mother of god and other essays* (pp. 310–319). University of Wisconsin Press.

Leopold, A. (1949). *A sand County Almanac, and sketches here and there.* Oxford University Press.

Lovelock, J. (2000). *Gaia: A new look at life on earth.* Oxford University Press.

Lynch, J. M., & Audus, L. J. (1976). Products of soil microorganisms in relation to plant growth. *CRC Critical Reviews in Microbiology, 5*(1), 67–107.

Maaz, T., Heck, R., Glazer, C., Lo, M., Zayas, J., Krenz, A., Beckstrom, T., Crow, S., & Deenik, J. (2023). Measuring the immeasurable: A structural equation modeling approach to assessing soil health. *Science of the Total Environment, 870*, 161900.

McClure, R. S., Farris, Y., Danczak, R. E., Nelson, W. C., Song, H.-S., Kessell, A., Lee, J.-Y., Couvillion, S. P., Henry, C., Jansson, J. K., & Hofmockel, K. (2022). Interaction networks are driven by community-responsive phenotypes in a chitin-degrading consortium of soil microbes. *mSystems, 7*(5), e00372-22.

Millstein, R. (2024). *The land is our community.* University of Chicago Press.

Natural Resources Conservation Service (NRCS). (2024). *Soil health.* Retrieved on 2 March 2025 from https://www.nrcs.usda.gov/conservation-basics/natural-resource-concerns/soils/soil-health

O'Neill, R. (2001). Is it time to bury the ecosystem concept (with full military honors, of course!). *Ecology, 82*(12), 3275–3284.

Powlson, D. S. (2020). Soil health—Useful terminology for communication or meaningless concept? Or both?' *Frontiers of Agricultural Science and Engineering.* https://doi.org/10.15302/J-FASE-2020326.

Powlson, D. S., Cai, Z., & Lemanceau, P. (2015). Soil carbon dynamics and nutrient cycling. In E. Milne, E. Noellemeyer, & S. Banwart (Eds.), *Soil carbon: Science, management and policy for multiple benefits* (SCOPE Vol. 71, pp. 98–107). CABI International.

Quastel, J. H. (1954). Leeuwenhoek lecture: Soil metabolism. *Proceedings of the Royal Society B, 143*(911), 159–178.

Roesch, L. F., Fulthorpe, R. R., Riva, A., Casella, G., Hadwin, A. K. M., Kent, A. D., Daroub, S. H., Camargo, F. A. O., Farmerie, W. G., & Triplett, E. W. (2007). Pyrosequencing enumerates and contrasts soil microbial diversity. *The ISME Journal, 1*, 283–290. https://doi.org/10.1038/ismej.2007.53

Singh, P., Kawa, N. C., & Sprunger, C. D. (2023). More questions than answers: Ohio farmers' perceptions of novel soil health data and their utility for on-farm management. *Agroecology and Sustainable Food Systems.* https://doi.org/10.1080/21683565.2023.2270928

Vargas, R. (2022). Soils, where food begins. *United Nations Chronicle.* Retrieved on 1 January 2025 from https://www.un.org/en/un-chronicle/soils-where-food-begins

Wade, J., Beetstra, M. A., Hamilton, M. L., Culman, S. W., & Margenot, A. J. (2021). Soil health conceptualization differs across key stakeholder groups in the Midwest. *Journal of Soil and Water Conservation*, 1–7. https://doi.org/10.2489/jswc.2021.02158

Wrightson, J., & Newsham, J. C. (1919). *Agriculture, theoretical and practical.* Appleton.

Chapter 3
Economics in the Big Wide World

Nancy Cartwright

Abstract John Dupré notoriously questions 'Could there be a science of economics?' He tells us his skepticism is directed especially to the mathematical constructs in economics, adding that '…there may well be much useful, if generally loose, causal knowledge to be had in economics….' This paper agrees with Dupré on just that. The exact conclusions that economics is so proud of hold only in what I describe as 'small worlds': settings where the only causes of the phenomenon of interest are ones that economics knows how to model. But most real-world settings aren't like that. Following Dupré's lead on 'loose knowledge', I argue that, in order to help with the large world outside its borders, economics should export the rigorously-established principles of which she is so proud as *qualitative, inexact tendency principles.*

3.1 Introduction

In 1993 John Dupre asked, 'Could there be a science of economics?' His answer: (Dupré, 1993a, 1993b, 1993c, 363–378.)

> A qualified 'Very probably yes'. First, my skepticism has been directed to theoretical, especially mathematical constructs in economics. I do not deny that there may well be much useful, if generally *loose*, causal knowledge to be had in economics…. (Dupré, 1993a, 1993b, 1993c, 377)

I agree. In this paper I will argue that economics, which aims to be an exact science, can be so. But only in 'small worlds'. And most real-world settings are not like small worlds. As John says, 'The complexity of the phenomena investigated by economics

N. Cartwright (✉)
Durham University, Durham, England
e-mail: nancy.cartwright@durham.ac.uk

University of Johannesburg, Johannesburg, South Africa

University of California, San Diego, California, USA

S. Leonelli et al. (eds.), *Life Beyond Human Boundaries: John Dupré's Philosophy and the Interdisciplinary Study of the Life Sciences*,
https://doi.org/10.1007/978-3-032-26977-5_3

is such as to make claims of fundamental lack of order at least superficially plausible.' (Dupré, 1993a, 1993b, 1993c, 367).

What then should we do? My answer is a development of John's original idea of **'loose… knowledge'.** (Dupré, 1993a, 1993b, 1993c, 377).

3.2 Economics as Exact

Economics prides itself on being exact and rigorous. For instance in a paper titled *The Superiority of Economists,* the historian/sociologist of economics Marion Fourcade and associates quote 'an eminent economics professor' as explaining:

> You are only supposed to follow certain *rules*. If you don't follow certain rules, you are not an economist. So that means you should *derive* the way people behave from *strict* maximisation theory… The opposite [to being axiomatic] would be arguing by example. You're not allowed to do that … People say 'that's anecdotal.' That's the end of you if people have said you're anecdotal… [T]he modern thing [people say] is: 'it's not identified.' God, when your causality is not identified, that's the end of you. [All ital is added] (Fourcade et al., 2015, 92)

The Harvard economist Dani Rodrick agrees with this description in his 2015 book, *Economics Rules: the rights and wrongs of the dismal science.* He tell us: 'Once a model is stated in mathematical form, what it says or does is obvious to all who can read it. This clarity is of great value.' (Rodrik, 2015a, 2015b, 23). The same thing is echoed by Jonathan Gruber in his undergraduate Microeconomics course at MIT:

> What makes economics great, and quite frankly what makes us the highest-paid of the social sciences, is that we can, with very few pretty basic assumptions, explain a wide variety of phenomena in the real world. […] *We're never going to get it perfectly* (ital added), but you'll be amazed at how the small a number of assumptions we need to explain an enormous number of things. (Transcribed from Jonathan Gruber's undergraduate Microeconomics course at MIT)

What accounts for economists' assumption that this way of proceeding is going to work? Economic historian and biographer of John Meynard Keynes, Robert Skidelsky, in his book *What's Wrong with Economics?* attributes this insistence on rigour and exactness in part to the conviction that, despite all appearances to the contrary, economists operate from 'the conviction that underneath the messiness of appearances there exists an underlying order that can be captured by logic and mathematics'. (Skidelsky, 2018, 28).

3.3 Problems

This gives rise to two interrelated problems. The drive for exactness pushes economics.

A. To study small worlds

B. To treat real settings as small worlds

And anyway, I shall argue, economics is *not* best practiced as an exact science.

Consider A. first: The drive for exactness pushes economics to study small worlds. My claim, recall, is that economics can be exact in small worlds and only in small worlds. Consider how physics predicts precise results. Physics establishes sets of principles meant to be universal. When it uses these principles for precise prediction or precise control of settings in the real world, the results are derived within models where, unsurprisingly, all the relevant factors for the phenomenon of interest are ones physics knows how to treat precisely. Otherwise it cannot yield precise results. That reflects the well-known principle: *Waffle in*; *waffle out*.

So physics' precise successes are in what I call 'small worlds'.

But what exactly is a small world? 'Small world' is a description meant to apply to real settings in the real world. It is a relative description—relative to a given phenomenon and a given body of knowledge or discipline:

> A real setting S is a *small world* for a phenomenon X for a body of knowledge K if and only if

1. All the factors in S that affect X are ones that K knows how to treat.
2. K knows how to calculate what happens when these factors occur together.

This view about physics is nicely reflected in an observation that the originator of econometrics, Trygve Haavelmo, once made to me in conversation: No-one asks physics to predict the course of an avalanche. But economists are asked to predict the course of the (avalanche of the) economy.

Turn now to B: The drive for exactness pushes economics to treat real settings as small worlds. Here is what Skidelsky, has to say about this:

> Physics has in nature its own ready-made laboratory, where events regularly repeat themselves. The social world lacks such stationary features. The standard economic model is typically a theoretical representation of a closed system. (Skidelsky, 62)

> What economics does is convert open systems into closed systems by excluding 'moves' that would render the system unstable. Dictators 'freeze the frame' by order: economists do it by 'modeling'. (Skidelsky, 2018, 3)

> Nobel Laureate Thomas Sargent (b.1943) defines a person as a 'constrained, intertemporal, stochastic optimising problem'. (Skidelsky, 2018, 81)

> Expectations about the future are simply probability distributions over a sequence of events. Uncertainty is reduced to probability, and can thus be labelled a special case of certainty. (Skidelsky, 2018, 81)

> The neoclassical model of rationality which Kahneman and Tversky set up as their benchmark [to challenge] might make sense in a small, closed world with well-defined limits. The coin-toss experiment is supposed to replicate this—it is either a head or a tail—but it is irrelevant as a test of rationality in open systems admitting of many different outcomes. (Skidelsky, 2018, 90)

This is similar to the view of two other leading economists, John Kay, who is well-known outside economics because of his weekly column in the *Financial Times,* and

Mervyn King, also more widely known because he served as Governor of the Bank of England:

> Asked: "What insights do we gain from this model?' The author's answer: 'The numbers derived from the model should be the policy'. That answer begins to make sense only if the economist believed that his model described 'the world as it really is'. (Kay & King, 2020. 261)

But a small world model does not do that.

You may be tempted to ask Skidelsky, 'What's wrong with converting open systems into closed systems? With 'freezing the frame' by 'modeling'?' The answer is that generally *it just doesn't work*. Significant relevant features studied in other fields can't be properly fitted into an economics frame.

Of course there are a variety of ways to cheat. You can get a desired outcome from an econometric equation by putting in an exogenous shock of just the right size. Physics does this with forces all the time. And in both fields it is ad hoc! Or you might assign utilities arbitrarily in a way that gets the right answer. This too is ad hoc! Or gerrymander choice sets, preferences, probabilities so that actions observed are in accord with maximising expected utilities. Again, this is ad hoc! Or you might choose weightings in measures and models that produce just the right outcomes. Yet again, ad hoc! Etc.

These all amount to giving up on exact science whilst preserving the pretence of it.

From which I conclude: economics must work in the big wide world. And it must do so together with others. If it is to be of genuine help, it must muck in with a motley assembly of other knowledge sources, a claim that is echoed by Skidelsky:

> … [economics] needs other fields of study, notably, psychology, sociology, politics, ethics, history to supply the gaps in its method of understanding reality. We should not be afraid to say to the economist, "There are more things in heaven and earth, Horatio, than are dreamt of in your philosophy". (Skidelsky, 2018, 78)

3.4 Solutions

There are two opposing strategies for dealing with these problems:

1. Bring lessons from other knowledge sources into economics.
2. Bring lessons from economics out.

Look at 1. first: Bring lessons from other knowledge sources into economics. Here are some examples of this strategy. You might pretend that all other sources of influence on decisions are genuine utilities. And you might do so despite Amartya Sen's strong arguments to the contrary in his criticisms of rational choice theory[1] and his insistence that we also act from commitment. (Note though that it is not just economists. Other fields do the same kind of thing. For example Michelle Foucault sees everything as

[1] Sen (1982), Sen (2002).

a matter of power relations.) A second strategy is to treat all causes as quantities, willy-nilly, and then add them into economic models. A third strategy that spread for a while from economics into other social sciences is to model everything as a game. And so forth. There is a short verdict to be delivered on all of these. They are a bad idea! They too often result in a false sense of security and failed prediction and explanation.

Turn now to 2: Bring lessons from economics out. The good news here is that economics can contribute this way. But the bad news is that to do so it seems she must often compromise her most prized virtues: rigour, exactness and subject- and value-free objectivity (which, as is widely argued nowadays, she probably never had in the first place).

On the good news side, I want to suggest both that what she can offer is a great help with prediction, albeit qualitative and that the vaunted rigour and exactness inside economics make her qualitative, inexact contributions outside exceptionally useful. And also we should keep in mind that in general sacrificing some rigour can be okay because, as my late husband, the philosopher Stuart Hampshire, urged: we should not bring more precision to a topic than it can bear.

This brings me to my central thesis:

> In order to help with the large world outside its borders, economics should forego its pride in being entirely quantitative and exact. In particular economics should export her rigorously established principles as qualitative, inexact *tendency principles*.

These can be especially useful because their rigorous backing in economics models can show what kinds of settings they can obtain in.

3.5 Tendency Principles

What though are tendency principles? This is a concept that so far as I know was first introduced by JS Mill in his arguments that economics is an analytic science.[2] It analyses a situation into its relevant causes, figures out what effect each of those causes *tends* to produce and then calculates what happens when all these tendencies contribute together. So tendency principles say not what happens when a factor is in play but what it tends to contribute.

Tendency principles have several important features. First, they can be qualitative. They often register no quantitative measure of strength, though sometimes they may provide a rough indicator, such as strong / weak and widespread / less common. Second, tendencies that obtain in a setting often pull in different directions. Clearly this is the case when different features tend to produce different outcomes, as when gravity pulls a metal object down but a magnet holds it up. But also, one and the same feature can have two opposing tendencies at the same time. Consider the dilemma about the effects on uptake of selling malarial bed nets at low cost versus giving

[2] Mill 1848

them away for free. Selling them has a tendency to induce people to value them and thus it can increase uptake, but it also increases the cost, which tends to decrease uptake. Third, there may be no stable rule of combination. This is in contrast with, for example, the tendencies associated with different forces in classical mechanics, where we calculate what happens when a number act at once by vector addition of the separate contributions. Fourth, tendency principles are often expressed as generics. They may have no quantifiers. Their range of possible application need not at all be universal. In fact it is often only implicit what the range is—and often not even that.

The fact that economics trades in tendencies may be concealed by talking instead about 'economic mechanisms'—which when spelled out as propositions would naturally be rendered as generic tendency principles. Here are a number of widely recognised economic mechanisms:

Opportunity cost
Marginal vs average cost
The whole standard setup of consumer theory

diminishing marginal utility
indifference curves
substitution and income effects

Externalities
Free riding
Monopoly / monopsony pricing, power, rents.

Many economists clearly recognise this. For instance, Skidelsky is explicit in endorsing the idea of economic principes as statements of tendency:

> The human apple may have a tendency to fall to the ground, but this tendency is far too weak to be called a law. There is nothing in the 'market' akin to the law of gravity.*(*Skidelsky, 2018*)*

For more examples, look at the form of some of the propositions that a very recent article[3] reports had strong agreement among economists each time polled:

Universal health insurance coverage will increase economic welfare in the United States.
Addressing biases in individuals and institutions can improve both equity and efficiency.
Climate change poses a major risk to the US economy.
Tariffs and import quotas usually reduce general economic welfare.
Flexible and floating exchange rates offer an effective international monetary arrangement.

And here's a citation of the opposing tendencies I noted with respect to selling bednets, from an article titled 'Free Distribution or Cost-Sharing: Evidence from a Malarial Prevention Experiment in Kenya'[4]: 'Some argue that those who pay

---

[3] Geide-Stevenson and La Parra-Pérez (2024).

[4] Jessica Cohen and Dupas (2014).

for goodwill value it more....Charging for nets does however considerably reduce access.'

The importance of tendency principles with no explicit quantifiers and often no exact quantities is that they can travel far more readily from economics to be used in other domains. This is by contrast with more exact principles—ones that are properly quantified and commit to precise strengths of influence. These precise numbers are hard to duplicate in the big wide world, whereas the general tendency may operate in a variety of settings despite their messiness. Indeed, the use of economic concepts without presumption of quantification or an optimality benchmark is widespread in political science and sociology. For instance you can find it throughout the comparative capitalisms field (see for example Hall & Soskice, *Varieties of Capitalism*) and in economic sociology (see for example *The Socio-Economic Review*).

So we have arrived at one of my central conclusions: economics can export hard-won lessons as qualitative tendency principles. But is this damming with faint praise? I say 'No'.

That's for two reasons. First, this is often as good as you can get—and it can be very useful for prediction, ala Mill. Second, the rigorous derivations of the more exact versions of these principes that are characteristic of economics provide guidance on application outside, guidance that's generally otherwise unavailable.

3.6 Guides to Range of Application

It's not just me that sees the export of economics principles this way. Here is the economist Dani Rodrick again:

> ....models contain information about the circumstances in which they are relevant and applicable. They tell us when we can use them, and when we might not...[They] come with explicit user's guides--teaching notes on how to apply them. That's because they are transparent about their critical assumptions and behavioral mechanisms. [Rodrik., 2015a, 2015b, 43]

Let's look at a specific example to get a sense of how this works. I choose Robert Lucas's 1976 model of the Phillips curve since it is very well known (not because I think it gets its overall lesson right). The Phillips curve can be expressed as a tendency principle: rising inflation tends to drive unemployment down (if only in the short term). Lucas' model is a prime example of exact economics. He *derives* the principle from a set of assumptions. But the central lesson Lucas draws is about the range of application of the principle. The model is standardly taken to show that the inflation-unemployment trade-off occurs *only where* employers mistake inflation for price rises in their sector.

The model makes a number of theoretical assumptions:

There is only type of shock: monetary
Information is imperfect
Agents are price takers
Agents' decisions are driven by relative (market-specific) prices

Ideal competition
Markets-clear
Agents maximize their expected utilities under constraints using all information available

And, as Skidelsky notes: 'In this model there is no attempt to investigate changes in institutional facts (in trade union organisation, historic levels of unemployment, among others) which might explain the breakdown of the original Phillips curve'. (Skidelsky, 2018, 65). This is a reminder that what the model studies is the tendency of inflation to reduce unemployment if, but only if, it is not recognised as inflation that is due to the operation of utility maximisation. So (supposing the model results are right), you can expect this tendency to be present when—but only when—utility maximisation is operating, and in ny one of those cases you can expect what happens to be a joint outcome of all the tendncies at work together in that case.

From these Lucas derives the trade-off—and he derives that it will occur only so long as employers mistake inflation for a rise in the price of the products they produce in their sector, not a rise in price in general. The idea is that when employers think they can sell their products at higher prices, they will want to make more of them and this will induce them to hire workers to produce these products. But naturally not if they recognise that all prices are going up, including the prices of the products they need to buy in order to produce the ones they sell.

Note that this is at best a tendency principle, with the characteristic features of a tendency principle. In particular, as Skidelsky notes: 'In this model there is no attempt to investigate changes in institutional facts (in trade union organisation, historic levels of unemployment, among others) which might explain the breakdown of the original Phillips curve'. (Skidelsky, 2018, 65). This is a reminder that what the model studies is the tendency of inflation to reduce unemployment if, but only if, it is not recognised as inflation that is due to the operation of utility maximisation. So (supposing the model results are right), you can expect this tendency to be present when—but only when—utility maximisation is operating, and in any one of those cases where it is present you can expect what happens to be a joint outcome of all the tendencies at work together in that case.

The point here is that the model provides a *derivation* from a set of explicit assumptions. So the Phillips-curve tendency must obtain where those assumptions are satisfied.[5] It occurs under these circumstances only as a tendency of course, since, as Skidelsky explains, many other tendencies may affect unemployment as well. Nevertheless, this may well be a whole lot more than other social science can offer on the subject. We are offered both a tendency and a range (albeit possibly a narrow one[6]) across which we can be sure that tendency will be present. Of course it may well operate more widely. But at least we can be assured it will operate over the range identified by the assumptions of the model.

[5] Supposing that idealisations—e.g. representative agents—don't make a difference.

[6] The range might well be extended though since economist are big on relaxing the assumptions in their models to extend the range of applicability of their conclusions.

3.7 My Overall Lesson then…

Wilhelm Lexis criticised Carl Menger's programme for only being able to offer general tendencies not accurately usable for precise prediction. The lesson I draw from this discussion is that Lexis was wrong. Tendencies really matter to understanding the large messy world where economics can help. But precise prediction is generally not possible. So.

> Don't be afraid of exporting the qualitative and inexact—so long as it is well-grounded in economics itself.

References

Cohen, J., & Dupas, P. (2014). Free distribution or cost-sharing? Evidence from a Randomized Malaria prevention experiment. *IPA*, Sept. 02, 2014. https://poverty-action.org/free-distribution-or-cost-sharing-evidence-malaria-prevention-experiment-kenya#:~:text=Overall%2C%20given%20the%20large%20benefit,%2Deffective%20than%20cost%2Dsharing
Dupré, J. (1993a). Could there be a science of economics? *Midwest Studies in Philosophy, 18*, 363–378.
Dupré, J. (1993b). The complexity of the phenomena investigated by economics is such as to make claims of fundamental lack of order at least superficially Plausible. *Midwest Studies in Philosophy, 18*, 367.
Dupré, J. (1993c). Very probably yes. *Midwest Studies in Philosophy, 18*, 377.
Fourcade, M., Ollion, E., & Algan, Y. (2015). The superiority of economists. *Journal of Economic Perspectives, 29*(1) (Winter 2015), 89–114.
Geide-Stevenson, D., & La Parra-Pérez, Á. (2024). Consensus among economists 2020—A sharpening of the picture. *The Journal of Economic Education, 55*(4), 461–478. https://doi.org/10.1080/00220485.2024.2386328
Kay, J., & King, M. (2020). *Radical uncertainty: Decision-making beyond the numbers*. W.W. Norton & Company.
Mill, J. S. (1848). *Principles of political economy*. Oxford University Press.
Rodrik, D. (2015a). *Economics rules: The rights and wrongs of the dismal science*. W.W. Norton & Company.
Rodrik, D. (2015b). *Economics rules: The rights and wrongs of the dismal science* (p. 43). W.W. Norton & Company.
Sargent, T. (2015). It's not identified. God, when your causality is not identified, that's the end of you. In *The superiority of economists*, by Marion Fourcade, Etienne Ollion, and Yann Algan, 93. *Journal of Economic Perspectives, 29*(1), (Winter 2015), 89–114.
Sen, A. (1982). *Choice, welfare and measurement*. Blackwell.
Sen, A. (2002). *Rationality and freedom*. Harvard University Press.
Skidelsky, R. (2018). *What's wrong with economics?* Yale University Press.

Chapter 4
"And the Wine is Bottled Poetry": Dupré on Reductionism and the Mental

Adrian Haddock

Abstract This essay has two main agenda. The first is to bring out (what I take to be) the fundamental insight of chapter seven of John Dupré's book *The Disorder of Things* ("Reductionism and the Mental"). The second is to suggest that this insight contradicts a central commitment of Dupré's philosophy as a whole.

*

1. Stanley Cavell once spoke of the sense of "rightness and relief" that he experienced, as a graduate student, on discovering the work of J.L. Austin (Cavell, 1965, p. 214). I had a similar experience when I discovered the work of John Dupré. Here was a philosopher who spoke with a human voice, and who sought to resist the impoverished conception of the human promulgated by reductionists of various stripes—in economics, in psychology, and in philosophy. One of Dupré's most memorable broadsides against this tendency is the chapter of *The Disorder of Things* entitled "Reductionism and the Mental" (Dupré, 1995, Chap. 7).[1] My central aim in this essay is to bring out what I take to be the fundamental insight in this chapter, and to show how this insight is devastating for the very idea of reductionism about the mind. But I also want to suggest that this insight contradicts a central commitment of Dupré's philosophy as a whole—what Dupré calls his "empiricism".

2. It helps to begin with "classical reductionism" (p. 88). At the core of classical reductionism is a certain conception of reality, and of science's relation to reality.

[1] All unattributed page references in the text are to this volume.

I would like to thank Jane Calvert, Andrea Kern, and Philip Kitcher for very helpful comments and discussion.

A. Haddock (✉)
Universität Leipzig, Leipzig, Germany
e-mail: adrian.haddock@uni-leipzig.de

S. Leonelli et al. (eds.), *Life Beyond Human Boundaries: John Dupré's Philosophy and the Interdisciplinary Study of the Life Sciences*,
https://doi.org/10.1007/978-3-032-26977-5_4

Science—that is, empirical science[2]—consists in a hierarchy of different languages, each of which contains general terms which distinguish kinds of things. On the standard way of fleshing this out, at the lowest level is the language of physics, then that of chemistry, then that of biology, then that of psychology, and then (at the highest level) that of the social sciences. But for my purposes, what matters is merely the conception of a hierarchy of languages, at one of the higher levels of which is the language of psychology, or (equivalently) the language of the mind. This conception is not unique to classical reductionism: it is shared by many of its opponents. The mark of classical reductionism is that it adds to this conception the claim that the kinds that are distinguished by each language are, in each case, identical to the kinds that are distinguished by the language at the lowest level.

3. Classical reductionism often comes with a familiar qualification: the identities between the kinds distinguished by the language of the mind, and the kinds distinguished by the lowest-level language, may themselves differ across different "kinds of mentally capable creatures" (p. 147). In the familiar illustration: the kind of neural state with which pain is identical in human beings may not be the kind of neural state with which it is identical in rats (for example). This is often put by saying that mental states are "multiply realizable": the same kind of mental state may be "realized" by different kinds of lowest-level states in different "kinds of mentally capable creatures". Dupré targets this qualified version of classical reductionism no less than the unqualified version. He is explicit that his fundamental concern is the specifically human mind.

4. Dupré targets not merely classical reductionism about the (human) mind, but equally the position known as "eliminativism". These positions share a common assumption: if the language of the mind distinguishes any kinds, then these are identical to the lowest-level kinds. For this reason, both positions count, by Dupré's lights, as versions of reductionism. But whereas classical reductionism insists that the language of the mind distinguishes kinds that are identical to the lowest-level kinds, eliminativism denies this, and so—in the light of the common assumption—concludes that this language does not distinguish any kinds at all: the general terms of the language of the mind are, in this sense, empty. Eliminativists think of each language on the hierarchy as expressive of a theory, in the sense of a body of empirical beliefs: as they reject the analytic-synthetic distinction, they do not distinguish between those sentences of a language that merely spell out the meanings of terms and those that express empirical beliefs—it is empirical beliefs all the way down. They call the putative theory expressed by the language of the mind "folk psychology" because they think of it as the putative "theory of mind" employed by the benighted "folk". And they take the sentences that make up this putative theory to be, not merely false, but empty, on account of the emptiness of their general terms. As a result, they think that this language does not belong on the hierarchy at all: the putative theory which it expresses is to be supplanted by a genuine theory couched

[2] In Germany at least, this is no pleonasm.

in a language which does admit of reduction. (Their favoured candidate is some kind of not-yet-actual neuroscience.)

5. About this position, Dupré has strong things to say. He observes that "folk psychology provides us with the entire resources of language for making subtle distinctions between an indefinite array of possible mental states" (p. 158). Making distinctions of this kind is

> one of the main things that language is for, and one of the things that language users, especially the most talented users, are most skilled at applying it to. One should be astounded and even appalled at the scientistic arrogance that supposes that the dry and highly specialised terminology of an esoteric subfield of science should supplant the instrument developed over many centuries by the efforts of Shakespeare, Dante, and Dostoevsky—to speak only of the West—and many millions of others to describe the subtleties of the human mind. (p. 158)

Here we find Dupré's insight: folk psychology, and so the language of the mind, "provides us with the entire resources of language". Someone who was equipped with the entire resources of the language of the mind—someone who understood this language, in its entirety—would not merely be provided with the entire resources of one of the languages on the hierarchy as against others; they would be provided with the entire resources of language—they would understand language, in its entirety.

6. This is enough to raise the question whether the language of the mind is one of the languages on the hierarchy. If being equipped with its resources suffices to provide us with the entire resources of language, must it not contain the entire hierarchy?

7. Dupré's insight tallies with a certain familiar point. The language of the mind distinguishes not only general types of mental states—e.g., beliefs, and experiences—but also more specific types of mental states of these general types. To focus on belief: it distinguishes, e.g., my belief that snow is white; my belief that grass is green; my belief that there are no chairs in this room; my belief that there are chairs in that room; and so on. I could go on by giving as many examples as there are sentences of language. Dupré himself arrives at his insight by considering distinctions, not between types of belief, but between types of experience, and specifically, between types of experience of wine. But the point is the same. The language I use to distinguish types of belief reflects the language I use to distinguish states of the world; the language I use to distinguish types of experience of wine is merely a specific case of this, because it reflects the language that I use to distinguish types of wine—wines that are "romantic, banal, presumptuous, adventurous, and so on" (158). The language that Dupré uses here is somewhat Thurberesque—and this is reflected in his citation from Robert Louis Stevenson, displayed at the entrance to Napa Valley, which I have taken as my title: "And the wine is bottled poetry". But the important thing is not the rhapsodic character of the language but the fact that the language which is used to distinguish types of experience of wine reflects the language that is used to distinguish types of wine: that the complexity of the *mind* reflects the complexity of the *wine*, and more generally the complexity of the world. The distinctions drawn by the language of the mind reflect the distinctions which are drawn by the entirety of language. This is a familiar point, and it offers a way to begin to

appreciate Dupré's insight. But to understand the truth and the force of his insight, we need to see how his insight goes beyond this point.

8. To see this, let us focus on belief, and specifically on reports of belief in the first person, through (putative) sentences of the (putative) form "I believe that *p*". On the face of it, these are sentences which contain other sentences. To take a familiar example: "I believe that the Earth moves" is a sentence which contains the sentence "The Earth moves". But then there is a notorious problem. Here is a principle: replacing a referring term in a sentence (or its utterance) with a term that refers to the same thing cannot affect the truth value of the sentence (or the utterance). That principle has been held by some to define the very idea of a referring term.[3] But it seems that replacing "the Earth" in an utterance of "I believe that the Earth moves" with a different term that refers to the same thing can affect the truth value of the utterance. Suppose it were replaced with "the planet most often referred to in discussions of psychological *oratio obliqua*", for example. Surely someone's utterance of the sentence resulting from this replacement might be false, even though their utterance of "I believe that the Earth moves" is true? Frege (1948) was famously unwilling to accept exceptions to the foregoing principle, and as a result held that singular terms which refer to the same thing when they occur outside content-specifying that-clauses refer to different things when they occur inside these clauses. Any referring term refers in a certain way: what it refers to is its referent, and the way it refers is its sense. On Frege's view, in an utterance of "I believe that the Earth moves", the singular term "the Earth" refers, not to the planet to which it refers in the self-standing sentence "The Earth moves", but to its sense in that sentence. The same holds, *mutatis mutandis*, for every singular term. We can make this vivid by counting singular terms with distinct referents as distinct singular terms. On this way of counting, the singular term in subject-position in "The Earth moves" is not the singular term immediately following "that" in "I believe that the Earth moves". We could mark this by writing "I believe that the Earth* moves". The same problem arises, *mutatis mutandis*, for predicates. And Frege makes, *mutatis mutandis*, the same response. On his view, "moves", in an utterance of "I believe that the earth moves", equally refers to its sense in the self-standing sentence. So, given a related way of counting, and a related way of marking the point, we could write "I believe that the Earth* moves*". The effect is to understand, not just belief-reports, but pieces of *oratio obliqua* in general, as concerned with their own domain of objects, distinct from the ordinary objects with which the self-standing sentences associated with pieces of *oratio obliqua* are concerned.

9. It is tempting to think that Dupré's insight is a consequence of the familiar point from §7. Let us continue to focus on first-person belief-reports. Someone who understands the language of belief, particularised to these reports, in its entirety, thereby understands every replacement for "*p*", and thereby understands the entirety of language. But if Frege is right, then this last step does not follow. Someone who understands the sentence "I believe that the Earth moves" thereby grasps the senses

[3] See Quine (1986, p. 8).

of the referring terms contained in this sentence. But they do not thereby grasp the senses referred to by these referring terms. If we were to put what they understand by "the Earth" in this sentence into words, then we could not go beyond the definite description "the sense of 'the Earth' in a self-standing sentence". Someone who understands this description does not thereby understand the singular term that it concerns. Generalising the point: someone who understands the language of belief, particularised to these reports, in its entirety, thereby understands the bits of language which replace "*p*" in the reports, but does not thereby understand the entirety of language, because to understand these bits of language is only to understand a fragment of language as a whole. The way of counting, and the way of marking this, which we employed above, makes this vivid: someone who understands "I believe that the Earth* moves*" does not thereby understand "The Earth moves"; and the same holds, *mutatis mutandis*, for every first-person report of belief.

10. Frege's conception of these reports does not contradict the familiar point from §7. Even on Frege's conception, the reports reflect the language which distinguishes states of the world, in that the referring expressions contained in the bits of language following "that" in these reports refer to the senses of singular terms and predicates contained in sentences which distinguish states of the world. In this light, there can be no objection to the claim that, even on Frege's conception of these reports, someone who understands any sentence of the form "I believe that *p*" thereby understands that any sentence of this form is associated with a self-standing sentence of the form "*p*". But this does not mean that someone who understands any, or even all, sentences of the form "I believe that *p*" (or their utterances) thereby understands any self-standing sentence of the form "*p*". And this shows how Dupré's insight goes beyond the familiar point. It goes beyond it at least in that it rejects Frege's conception of reports of belief in the first person.

11. Of course, Frege's conception is far from being universally accepted. Donald Davidson rejects it in favour of his own influential "paratactic" account of oratio obliqua, for example (Davidson, 1968). But Davidson's account equally does not secure Dupré's insight. It holds that, in fact, there are no sentences of the form "I believe that *p*", but merely sentences of the form "I believe *that*", where "*that*" is a demonstrative referring to utterances of sentences. When we utter what seems to be a sentence of the form "I believe that *p*", we really utter two sentences: "I believe *that*", and "*p*". The first sentence has the significance of "I have a belief with the same content as that"; and the demonstrative in our utterance of this sentence refers to our utterance of the second sentence. The truth value of the first utterance can change depending on which utterance is demonstrated. So, a central feature of *oratio obliqua* is preserved. And the second sentence is not contained in the first sentence at all. So, the route to the notorious problem from §8, and so to Frege's conception of psychological *oratio obliqua*, is blocked. But Dupré's insight is not secured. On the contrary: first-person belief-reports consist in utterances of sentences which effect demonstrative reference to further utterances, and to understand the first array of utterances is not thereby to understand any utterance in the second array. The upshot is relevantly the same as on Frege's conception: someone who understands

the language of the mental, particularised to such reports, in its entirety, does not thereby understand the entirety of language.

12. Dupré's insight would be secured if, and only if, first-person belief-reports have the form of sentences containing sentences. Of course, that is the understanding of their form which generated the problem to which Frege and Davidson, in their different ways, respond. But this problem rests on the principle that replacing a referring term in a sentence (or its utterance) with a term with the same referent cannot affect the truth value of the sentence (or the utterance). It is not obvious that this principle must be accepted. Could there not be irreducibly non-extensional contexts which are not governed by this principle? Dupré's insight entails that first-person belief-reports violate the standard behaviour of *oratio obliqua*, given the principle. That leaves us with the question of whether to reject the principle for belief-reports *in toto*, or to claim an exception for those in the first person. We need not decide this here: it is enough to note that it is a question that Dupré's insight leaves us with.

13. What matters is that his insight has the following consequence. Return to the idea of the hierarchy of languages, each of which expresses a certain theory, in the sense of a certain body of empirical beliefs. Suppose that we are given the complete hierarchy of these languages, and so the complete expression of every theory. Every sentence, on every level of this hierarchy, can be accompanied by "I believe that...". The result, in every case, is a sentence, of the form "I believe that *p*", which contains the accompanied sentence. Suppose that we are now given the totality of the resulting sentences. Do we now have the expression of a further theory, over and above the expression of every theory? No—as we already had the expression of every theory, this totality perforce has a different status. At every level, the result of accompanying the sentences with "I believe that..." does not express a further theory. It expresses the same theory, but in a manner that makes explicit its internal character as a theory—that is, as a body of beliefs.

14. This is to understand "I believe that..." as having the same status as logical vocabulary, on a certain conception of what this vocabulary comes to. Consider the fundamental logical operations of conjunction ("... and __") and negation ("not-...").[4] The office of conjunction is not to contribute empirical information to the sentences it holds together, but merely to say that the things said by these sentences are true together, and in so doing, to make explicit that they are held together. Similarly, the office of negation is not to contribute empirical information to the sentences it negates, but merely to say that the things said by these sentences are not true, and in so doing, to make explicit that they are denied. In each case, the status of the piece of logical vocabulary reflects the fact that, if a theory did not already have the character which it is the office of the piece of logical vocabulary to make explicit, by saying what it does, then it would not be a theory at all. In the case of conjunction, it would lack unity: it would not be a body, but a miscellany of beliefs. In the case of negation,

[4] These operations are fundamental because every other truth-functional logical operation can be defined in their terms.

it would lack determinacy: its elements would not, in affirming one thing, be denying another, and as such they would not be affirming anything, and as such they would not be beliefs.[5] The same goes, *mutatis mutandis*, for "I believe that...". Its office is not to contribute empirical information to the sentences it accompanies, but merely to say that the things said by the sentences are things that the subject believes, and in so doing, to make explicit that they are beliefs. As with the two pieces of logical vocabulary, its status equally reflects the fact that, if a theory lacked the character which it is the office of "I believe that..." to make explicit, by saying what it does—if it were not a body of beliefs—then it would not be a theory at all.

15. Every pair of sentences, at every level of the hierarchy, can be accompanied by the first of these two pieces of logical vocabulary (i.e., by conjunction). This is not a contingency. The sentences are intrinsically such as to be so accompanied because conjunction merely makes explicit one of the internal characters of a theory—namely, its unity. We can put this by saying that, at every level of the hierarchy, the totality of sentences containing conjunction contains, and is contained in, the entire level. Indeed, we can go further: the totality of sentences containing conjunction contains, and is contained in, the entire hierarchy. Similar remarks apply, *mutatis mutandis*, to "not-...", and "I believe that...". In the case of the latter, if "*p*" is the entire hierarchy, then "I believe that *p*" contains, and is contained, in the entire hierarchy. And as such, it equally contains, and is contained in, every sentence, on every level of the hierarchy. In this sense, the language of the mind, particularised to first-person reports of belief, is not a language on the hierarchy at all.

16. That is Dupré's insight. And it is devastating for eliminativism and classical reductionism alike. Classical reductionists claim not to be eliminativists, on the grounds that, whereas eliminativists hold that the language of the mind is empty, they hold that it distinguishes kinds which are identical to the lowest-level kinds. But the truth is that classical reductionism is a form of eliminativism—because it shares the crucial assumption that, if the language of the mind is meaningful at all, then it is one of the languages on the hierarchy. If that assumption were true, then there would be no such thing as first-person reports of belief—because these reports belong to the language of the mind, and it is a condition of their meaningfulness that they are not on the hierarchy at all. That is what Dupré's insight lets us see. And it has the following consequence. First-person reports of beliefs are intrinsic to the language of the mind: if there was no such thing as these reports, then there would be no such thing as this language. So, because classical reductionism eliminates these reports—on account of the crucial assumption which it shares with eliminativism—it does implicitly what eliminativism does explicitly: it eliminates the language of the mind. And this consequence ramifies. If there was no such thing as the language of the mind, then the terms distinctive of this language, centrally "belief", and so "theory", would be meaningless. There would be no such thing as a belief, and so no such thing as a theory. The entire hierarchy would vanish. Eliminativism, in both its standard and

---

[5] At the most elementary level, this feature comes out in the equivalence of any of my beliefs with its double negation. In affirming that*p*, I deny that not-*p* (i.e., I affirm that not-not-*p*).

its classically reductionist versions, would not merely eliminate the language of the mind; it would eliminate every theory—not just "folk psychology" but neuroscience as well. If eliminativists do not see their position as having this consequence, then it is because of their assumption that, if the language of the mind is meaningful, then it is one of the languages on the hierarchy—for it is this assumption which makes it look as if the elimination of the language of the mind need not unsettle the rest of the hierarchy.[6] But this is the very assumption that guarantees the elimination of the entire hierarchy: the assumption that has this consequence equally blinds those who endorse it from seeing that it has this consequence.[7] This consequence is a catastrophe for eliminativism. Eliminativism, in both its standard and its classically reductionist versions is, or at least purports to be, a certain conception of reality, and of science's relation to reality. But if eliminativism were true, then there would be no such thing as science. If eliminativism were true, then eliminativism would be empty. In this way, eliminativism eliminates itself. That is a further consequence of Dupré's insight. It is worth noting that it is not merely the standard versions of eliminativism and classical reductionism which count as versions of eliminativism; the same would hold of a version of anti-reductionism which shares the assumption that, if the language of the mind is meaningful, then it is a language on the hierarchy, but insists that this language distinguishes kinds that are not identical to the lowest-level kinds. The shared assumption is enough to make this position no less self-eliminating.

*

17. Dupré's insight entails an influential idea in contemporary philosophy known as the "transparency of belief".[8] This treats an empirical belief as an answer to a question, of the form "Is it the case that *p*?", which is such as to be answered by actualising an epistemic procedure which intrinsically involves experience. An example would be a question which is such as to be answered by looking to see whether it is the case that *p*. Suppose that someone has answered a question of this kind by actualising such an epistemic procedure. And suppose that they go on to ask themselves a further question of the form "Do I believe that *p*?", where the belief, which forms the topic of this question, concerns the very same state of affairs which the former question concerns. What epistemic procedure do they need to actualise in order to answer this question? According to the idea known as the "transparency of belief", they do not need to actualise any epistemic procedure, over and above the one they have already actualised in answering the former question. That procedure delivers the empirical information which, on account of the empirical information which "*p*" contains, is needed to affirm "*p*" (or its negation). If "I believe that..." contributed further empirical information of its own, then actualising a further

[6] Or at least the rest of the hierarchy below the language of the mind.

[7] There is a voluminous literature defending eliminativism from various charges of self-refutation. Much of this literature is not without value because many of these charges were erroneous in letter. But they were sound in spirit: they merely failed to identify the fundamental assumption, through which eliminativism self-destructs.

[8] The idea is owed to Edgley (1969, p. 90), and to Evans (1982, p. 225).

epistemic procedure would be needed to affirm "I believe that *p*" (or its negation). But no further procedure is needed: the same procedure is sufficient to answer both questions—not because this procedure somehow delivers more empirical information than is strictly needed to answer the "world-directed" question, but because the answer to the "mind-directed" question does not contribute any empirical information of its own.

18. The "transparency of belief" is a consequence of the insight that first-person belief-reports contain, and are contained in, the entire hierarchy. That insight has the consequence that to accompany the sentences, at any level of the hierarchy, with "I believe that..." is merely to affirm the associated theory, but in a manner that makes explicit its character as a theory; it is not to express a belief about a different subject matter, as it would be if it were to affirm a different theory—one about "the mind", rather than "the world"—and as such, it does not involve any growth in empirical information.

19. Kant famously remarked that the "I think" must be able to accompany all my representations. I think that Dupré's insight is a way to understand this remark. But I do not think that his insight is uniquely Kantian. The idea of logical vocabulary as containing no empirical information is a central thought of the early Wittgenstein (1921).[9] And the later Wittgenstein's idea that "I believe" cannot be meaningfully said to figure in the expression of knowledge embodies a conception of this and other first-person phrases as incapable of being meaningfully said to make any contribution to the expression of empirical beliefs (Wittgenstein, 1958, p. 222).[10] Dupré tends to cite Wittgenstein as an influence,[11] and I am inclined to think that this influence is responsible for the passage in his writing where he expresses his insight.[12]

20. If empiricism is the idea that the meaningfulness of language consists in its containing empirical information, then Dupré's insight contradicts empiricism. That by itself is no criticism (on the contrary). But because Dupré is "a self-confessed empiricist" (p. 167), it might seem that he must endorse empiricism, so understood, and as such that there is a contradiction in his thought: between his insight and his empiricism. But when he speaks of his empiricism, he seems to have something

[9] It follows from the *Grundgedanke* of the *Tractatus Logico-Philosophicus* (4.0312).

[10] Cf. pp. 190–193, and §§404–409.

[11] For those only familiar with Dupré's work in the philosophy of biology, this influence might seem to belong to the same category as Sherlock Holmes's interest in the Polyphonic Motets of Lassus—a mere quirk of his personality, of no significance for understanding the fundamentals of his thought. But at the conference at which I read an early version of this essay, Dupré made clear that this was not his view.

[12] I cannot resist an anecdote here. In one of our supervision sessions in the early 2000s, John defended (and not merely for dialectical purposes) the later Wittgenstein's idea that the first-person statements "I believe that *p*", and "I am in pain" cannot meaningfully be said to express knowledge: although I *can say* "I believe that *p*" and "I am in pain", the things that I would thereby say are not things that I can meaningfully be said to know. (For a defence of this idea, for the case of "I am in pain", see Anscombe (1963, pp. 13–14).)

more subtle in mind—namely, that metaphysics, the attempt to articulate the fundamental form of reality, is an empirical enquiry, to be prosecuted by synthesising the results of empirical science. Perhaps there are discourses whose meaningfulness does not consist in their containing empirical information. Dupré's empiricism is that metaphysics is not a discourse of this kind.

21. But Dupré's insight contradicts this seemingly more subtle version of empiricism no less than the seemingly less subtle version. Our beliefs are transparent in two directions—to the world, and to themselves. Dupré's insight helps us to see this twofold transparency, and to see that these two directions belong together. Our beliefs are transparent to the world in that they are individuated by the world-directed sentences contained in first-person belief-reports. And they are transparent to themselves, in that their self-ascription is not a theory: no empirical inquiry is needed to move from an answer to the question "Is it the case that *p*?" to an answer to the question "Do I believe that *p*?" The non-empirical character of metaphysics is an immediate consequence of this twofold transparency.

22. We can see this as follows. The idea that our beliefs are transparent to themselves just is the idea that they are not merely conscious, but self-conscious. Their self-consciousness just is what is expressed by an answer to the question "Do I believe that *p*?" And, as we have seen, this answer contains no empirical information over and above the empirical information contained in an answer to the question "Is it the case that *p*?" In the self-consciousness of our beliefs, we are conscious of what we believe. And what we believe is not simply a plurality of elements, but a plurality of elements that hang together in a certain way, or (equivalently) a plurality of elements that bears a certain form. In being conscious of what we believe, we are conscious of what we believe as bearing this form. Fundamentally, we are conscious of it as bearing the form: *this is the case*. That is the fundamental form of what we believe. And it is equally the fundamental form of the states of the world. The sameness, of the fundamental form of what we believe and the fundamental form of reality, falls out of the fact that we distinguish specific types of belief by using the very language that we use to distinguish specific states of the world. The attempt to articulate the fundamental form of what we believe is a non-empirical inquiry, in that it is an attempt to articulate the fundamental form of what we are conscious of, in the self-consciousness of belief. And this just is the attempt to articulate the fundamental form of the states of the world. That is why metaphysics is a non-empirical inquiry: because it is the attempt to articulate the form that is (at once) the form of thought, and the form of reality. In spite of their differences, the metaphysical enquiries that we find in Aristotle's *Categories*, Kant's *Critique of Pure Reason*, and Hegel's *Science of Logic* all meet this abstract description: they seek to articulate the fundamental form of thought and reality and, in so doing, they seek to articulate the more determinate forms into which this fundamental form develops.[13] The idea that metaphysics is an empirical inquiry is nothing other than the idea that the articulation of what we

[13] For a contemporary presentation and exemplification of this understanding of metaphysics, to which I am deeply indebted, see Rödl (2012).

believe, and *a fortiori,* of the fundamental form of what we believe, is an empirical enquiry. And that is what Dupré's insight denies. When we think his insight through, we see that it is the negation of his empiricism.

23. Abandoning Dupré's empiricism would not unsettle his various attempts to synthesise the results of empirical (and specifically, biological) science into a comprehensive theory of the world. But it would mean that that is what it is: a *theory*, in the sense of the word that has been operative here—a body of empirical beliefs. It is not the sublime activity that characterises the philosophical tradition: the articulation of the fundamental form of reality.

24. The error, which underwrites reductionism about the mental, in all its versions, is empiricism about the language of the mind: the idea that the meaningfulness of this language consists in its containing empirical information. Dupré's insight contradicts empiricism, so understood, and in so doing contradicts his own empiricist conception of metaphysics. It equally sets us the important task of making sense of what it would be to reject empiricism about this language. Let me end by briefly saying something about this.

25. So far, we have focused on first-person reports of belief, and we have seen that they contain no empirical information, over and above that contained in the beliefs they report. But reports of belief are not only in the first person; they are also in the second and the third person. Are we to say the same about these? Or are we to tell a different story: that "You believe that *p*" and "They believe that *p*" do contribute empirical information of their own, and as such, that it is not true to say that they contain, and are contained in, what they report? If that were so, then they would be unified with first-person belief-reports in name only. Is there a way to avoid this, by making sense of second- and third-person belief-reports as equally containing no empirical information of their own?

26. A proper treatment of this question would require its own essay. But let me end with a suggestion. Just as saying "I believe that *p*" rests on no empirical procedure over and above that required to say "*p*", the same is true—in the explanatorily fundamental case by reference to which all others are to be understood—of the second- and third-person belief-reports. The only difference is that, in the case of the latter, the manner of answering the world-directed question, "Is it the case that *p*?", is an essentially social matter, in that it is a matter of actualizing the empirical procedure, not by oneself, but with others: one is in a position to ascribe the relevant belief, in the second and the third person, to others, on account of the fact that one actualizes the empirical procedure with them. These reports do not rest on the actualization of a further empirical procedure, over and above that required to answer the world-directed question—one directed, not at the topic of the world-directed question, but at the others. That would be to think of the reports as belonging, not to a study of this topic, but to a study of the others. And that is not how things are. In making these reports, one is engaged in a study of the topic—a study of the world—with others. Others figure here as those whom one studies *with*, rather than as those one makes a study *of*. That, I think, is a condition for understanding the real unity of

first-, second-, and third-person belief-reports, and so for a proper articulation of Dupre's insight. I cannot say more about "studying with" here. But it is the topic of Jane Calvert's essay, in this volume.

*

References

Anscombe, G. E. M. (1963). *Intention* (2nd edn.). Harvard University Press, 2000.

Cavell, S. (1965). Austin at criticism. *The Philosophical Review, 74*(2), 204–219.

Davidson, D. (1968). On Saying That. *Synthese, 19*(1/2), 130–146.

Dupré, J. (1995). *The Disorder of Things: Metaphysical Foundations of the Disunity of Science*. Harvard University Press.

Edgley, R. (1969). *Reason in Theory and Practice*. Oxford University Press.

Evans, G. (1982). *The Varieties of Reference*, ed. John McDowell. Oxford University Press.

Frege, G. (1948). Sense and Reference, trans. Max Black. *The Philosophical Review, 57(*3), 209–230.

Quine, W. V. O. (1986). *Philosophy of Logic* (2nd edn.). Harvard University Press.

Rödl, S. (2012). *Categories of the Temporal: An Enquiry into the Forms of the Finite Intellect*, trans. Sibylle Salewski. Harvard University Press.

Wittgenstein, L. (1921). *Tractatus Logico-Philosophicus*, trans. David Pears and Brian McGuiness. Routledge and Kegan Paul, 1961.

Wittgenstein, L. (1958). *Philosophical Investigations,* trans. G.E.M. Anscombe. Basil Blackwell.

Part II
Process Biology

Chapter 5
"Thou Hast Taught Me, Silent River, Many a Lesson, Deep and Long": Metabolism, Development, and Evolution as Team Sports

Scott F. Gilbert

Abstract We recognize three great fluxes in nature. First there is the metabolic flow of molecules that pervades our cells and which creates organisms. Second, there is the developmental flow of cells that is our cycle of life. Third, there is the evolutionary flow of genes, organisms, and populations from the first single-celled prokaryotes to the extensive biodiversity present on earth today. Recent studies have shown that organisms are consortia of several species acting together. What we had thought of as metabolism based on a single genome is actually the co-metabolism of more than one species. What we had thought as development from a single zygote is actually sympoiesis, co-development, where symbionts provide important signals for normal cell differentiation and maturation. And what we had thought as evolution is the co-evolution of several species together.

5.1 Replication vs Metabolism in the Theory of Life

5.1.1 Crystals

To Claude Bernard (1873/1974, p. 19), the founder of the cell physiology, any attempt to define "life" would be "stamped with sterility." Rather, "It is enough to agree on the word life to employ it; but above all it is necessary for us to know that it is illusory and chimerical and contrary to the very spirit of science to seek an absolute definition of it. We ought to concern ourselves only with establishing its characteristics and arranging them in their natural order of rank."

S. F. Gilbert (✉)
Department of Biology, Swarthmore College, Swarthmore, PA, USA
e-mail: sgilber1@swarthmore.edu

Facultyof Biochemistry and Molecular Medicine, University of Oulu, Oulu, Finland

S. Leonelli et al. (eds.), *Life Beyond Human Boundaries: John Dupré's Philosophy and the Interdisciplinary Study of the Life Sciences*,
https://doi.org/10.1007/978-3-032-26977-5_5

In 1982, I had proposed that there were two major traditions that had emerged to characterize living organisms: replication and metabolism. The tradition of seeing life as replication gave rise to molecular biology. The tradition of metabolism gave rise to biochemistry. They are currently reintegrating in several ways, and John Dupré has been instrumental in forging one of these syntheses, seeing even the replicative lineages as being subsumed in the patterns of metabolic flux.

The first tradition, that of replication, saw crystals as their metaphor and crystallization as a model for organismal growth. This model becomes important in the seventeenth and eighteenth centuries. For instance, in 1811, de la Métherie saw in crystallization the mechanism of organic growth and development: "The seeds of the male and of the female, being mixed, act as two salts would and the result is the crystallization of the fetus" (Ritterbush, 1968). This was, after all, the age of embryonic Preformationism. Such crystallization became a central portion of Theodor Schwann's cell theory, which hypothesized that new cell formation and growth could be explained by the same processes that govern the formation of inorganic crystals. "Organisms," wrote Schwann, "are nothing but the form under which substances capable of imbibition crystallize" (Schwann, 1839 quoted in Mendelsohn, 1965). In this view, the nucleus (cytoblast) would condense out of solution around the nucleolus and then the cell substances are deposited on the nucleus in the manner of a growing crystal (Florkin & Stotz, 1973). However, Schwann's ideas of cell continuity were disproven by new cell theories, such as those of Robert Remak, who demonstrated that new cells arose through the division of pre-existing cells.

Instead, the notion of crystallization became applied to chromosomes and genes (Gilbert, 1982). Haldane (1932, 156) was one of the first to propose that whatever genes may be, they would replicate as crystals.

> The growth and reproduction of large molecules are not, it may be remarked, hypothetical processes. They occur, it would seem, in certain polymerizations which are familiar to organic chemists. In my opinion, the genes in the nuclei of cells still double themselves in this way. The most familiar analogy to the process is crystallization. A crystal grows if placed in a supersaturated solution, but the precise arrangement of the molecules out of several possible arrangements depends on the arrangement found in the original crystal with which the solution is
>
> 'seeded.' The metaphor of seeding, used by chemists, points to an analogy with reproduction.

Joseph Needham took this metaphor and made it concrete. For him, crystals were not merely models or analogues for the living process, but are the physical components of cells. "Next, there is the profound importance of the paracrystalline state of biology. The 'liquid crystal' is not merely a model for what goes on in the living cell; it is in point of fact a state of organization actually found in the living cell" (Olby, 1971). This raises the stakes considerably. What is now being proposed is that the crystalline state is, in fact, found in the living cell.

A "proof of principle" for this was soon forthcoming, since the same year that Needham proposed these ideas, Wendell M. Stanley crystallized tobacco mosaic virus, a feat that was recognized as a philosophical breakthrough similar in its significance and interpretation to Wohler's synthesis of urea. Stating that this feat demonstrated "that there is no definite boundary between the living and the nonliving," the authors of one textbook (White et al., 1959, p. I) relate that "certain viruses have been isolated in highly purified crystalline form and studied extensively in the laboratory. Yet these substances, isolated in the laboratory and having no apparent or obvious features characteristic of living matter … an inanimate crystalline compound, when introduced into the proper environment, appears to behave as though it were a living viral agent."

Within a year, geneticist Hermann Muller claimed that this crystallized virus was apparently a pure protein capable of "autosynthesis" and that "it represents a certain type of gene" (Olby, 1971). At the Genetical Congress in Edinburgh in 1939, no less than three crystallographers, Astbury, McKinney, and Gowen, put forth analogies between viruses and genes. Astbury said that viruses and chromosomes were the two simplest reproductive systems known. Thus, autocatalysis, which was thought to occur by crystalline duplication, was now seen to occur in viruses which could themselves be crystallized. Furthermore, since this process is also common to chromosomes, the chromosomes, too, were thought to be crystalline in nature.

The virus and the chromosome—crystalline, nonmetabolizing, and capable of enormous feats of replication [see, e.g. (Monod, 1971, pp. 12–13, 18)]—became the unit of life for molecular biologists. Schroedinger's essay, "What Is Life?" claiming life to be an aperiodic crystal, influenced a generation of scientists, including F. Crick, J. Watson, M. Wilkins, S. Luria, S. Benzer, and even E. Chargaff. Yoxen (1979) has claims that it "exerted on some scientists a powerful and transient influence in suggesting and validating a particular line of research,", namely, that crystallinity is a state unifying all of matter and that "the most essential part of a living cell—the chromosome fiber—may suitably be called an aperiodic crystal." Indeed, prior to the discovery of DNA polymerase, it was thought that DNA might replicate itself autonomously as a crystal. It is interesting to read the last "coy" sentence of the classic Watson and Crick (1953a) *Nature* paper in light of crystal replication. It reads, "It has not escaped our notice that the specific pairing we have postulated immediately suggests a possible copying mechanism for the genetic material." In a second paper that year, Watson and Crick (1953b) hypothesized that the nitrogenous base precursors of DNA would line up ("crystallize") on their template and be zippered together. No enzyme was needed for DNA replication. Indeed, even after that discovery of DNA polymerases, the process was still described as crystalline replication (see Pfeiffer, 1960).

5.1.2 Whirlpools

Whereas molecular biology adopted the notions of replication and its model of crystals, biochemistry adopted the notion of metabolism and its model of whirlpools. The whirlpool/river model of life has an equally long history. The whirlpool has been seen as persisting, despite its constantly changing its components. Indeed, the whirlpool exists because it constantly changes its components (Gilbert, 1982). According to Claude Bernard (1873/1974), it was Baron von Cuvier who said, "The living being is a whirlpool constantly turning in the same direction, in which matter is less essential than form." Bernard agreed, adding that, "The particles of matter that enter the vital whirlpool are more complicated than those that emerge from it .... The energy set free in this fragmentation is the source of the active forces of the organism."

Bernard's view links the whirlpool model with chemical synthesis, and this was to be a crucial notion in biochemistry. Thomas Henry Huxley (Huxley, 1870, 402) also remarked that the constituents of an organism "stand to it in the relation of particles of water to a cascade, or a whirlpool'. Moreover, just as "the stoppage of a whirlpool destroys nothing but a form, and leaves the molecules of the water, with all their inherent activities intact, so what we call the death…of an animal, or of a plant, is merely the breaking up of the form, or manner of association, of its constituent organic molecules.'".

The whirlpool is still there for John Dupré's student, Dan Nicholson (2018), who claims "The existence of a whirlpool, for instance, is a direct consequence of its own unremitting activity, which is what enables it to maintain itself through time. In the same way, if an organism is to stay alive, it has to keep acting ('working', in the thermodynamic sense) to avoid the ever present threat of equilibrium." It is important that Nicholson links the metabolic view with thermodynamics. For, in such a view, thermodynamics allows attractor basins of local stability at energy levels far from final equilibrium. Indeed, metabolism becomes both the mechanism through which life is able to persist against the entropic demands of the universe.

This linkage of metabolism with thermodynamic attractor basins will have important consequences; for the molecules that enter our body do not stay the same. Rather, they will be transformed and utilized for the production of the energy and compartments needed to persist through their jiu-jitsu with the Second Law of Thermodynamics. Our contemporary view of metabolism was proposed in the 1940s as a result of the new ability to follow metabolic pathways with radioisotopes. Rudolf Schoenheimer was probably the first investigator to tag nutrients with radioactive chemicals and watch their transformation as they became different molecules and different body parts (Schoenheimer & Rittenberg, 1935, Schoenheimer, 1942; Guggenheim, 1991; Landecker, 2013). The current view of metabolism, wherein all the components of an organism are in a constant state of chemical renewal, was formulated by Schoenheimer, and these insights were formulated into metabolic pathways by him and by several other professors, many who had fled Hitler to come to Great Britain or the USA (Kennedy, 2001; Holmes, 1991, 512). Hans Krebs, Otto Meyerhoff, and Fritz Lipmann, for instance, helped elucidate the schema whereby oxygen, the byproduct of photosynthesis, served as the receptor for electrons from dietary nutrients, allowing cells to make the energy-storing ATP molecules that allowed life to persist against entropic gradients. As Schoenheimer (1942, p. 65) wrote,

> The new results suggest that all constituents of living matter, whether functional or structural, of simple or of complex constitution, are in a steady state of rapid flux. The finding of the rapid molecular regeneration...suggests that the biological system represents one great cycle of closely linked chemical reactions.

In 1966, Hans Jonas realized the philosophical importance of such a view. "The exchange of matter with the environment is not a peripheral activity engaged by a persistent core: it is the total mode of continuity (self-continuation) of the subject of life itself." Life, for Jonas, is the set of relationships wherein the only way to retain one's individuality is by changing one's parts. This is the meaning of metabolism. An organism "is never the same materially and yet persists as its same self, *by* not remaining the same matter." (Jonas, 1966, 76; see Gilbert, 1982; Landecker, 2017). Thus, the living organism is characterized by what Jonas (1966, p. 80) has called "a needful freedom," as metabolism constitutes both the identity of the organism and, simultaneously, its dependence on its environment. On one hand, "the organism must appear as a function of metabolism rather than metabolism as a function of the organism." But on the other hand, "we have to understand life not as a property of an already constituted living being, but as the act by which it constitutes itself and individuates." As Anne Sophie Meincke (2022) has expounded, this dialectic allows one to perceive an organism as an identity-maintaining individual unit as well as part of a homogeneous and non-individualistic world of metabolism. Indeed, "the very opposite of isolation emerges from the isolation of the organic self" (Jonas, 1966, 85). Life is where the Second Law meets the Third Critique.

Thus, we have two traditions in the life sciences for the chemical bases of life: replication and metabolism. How we view life depends on how we synthesize these two views. And here is where I wish to bring in the work of John Dupré.

5.2 Co-metabolism

John Dupré is fond of quoting Heraclitus: "No man ever steps in the same river twice, for it's not the same river and he's not the same man." So let me speak of a man and these rivers. There is, of course, the currents of a person's life, which can be translated into Latin as as one's "*curriculum vitae*." And this current contains three interacting streams that characterize life:

- *The flow of metabolism:* This is the flow of chemicals, mentioned above, that allows one to retain one's identity by changing one's parts.
- *The flow of development:* This is the life cycle through gamete formation, fertilization, cleavage, gastrulation, organogenesis, adulthood, and senescence.
- *The flow of evolution*: This is the flux of genes from one species to another, as life branches into new species.

I had claimed (Gilbert, 1982) that these two traditions of charactering life–replication and metabolism—had come together in the operon model of Jacob and Monod (where the production of a particular RNA depended upon which biochemical pathway was operating within the cell) and the semiconservative model of DNA replication (where an enzyme took instructions from its polynucleotide substrates). And in 2009, John Dupré and Maureen O'Malley wrote their seminal paper, "Varieties of Living Things: The Intersection of Lineage and Metabolism." Here lineage (the replication pathway) and metabolism (the pathway of flux) unite. In this paper, they get to the heart of the new view of organisms:

> What is a biological individual? Our central argument is that life arises when lineage-forming entities collaborate in metabolism. By conceiving of metabolism as a collaborative process performed by functional wholes, which are associations of a variety of lineage-forming entities, we avoid the standard tension between reproduction and metabolism in discussions of life.

Avoiding the "standard tension between reproduction and metabolism in discussions of life" means nothing less than that life is not merely metabolism or reproduction. Rather, evolutionarily separate replicative entities are joined into a "co-metabolism." In 2013, Smith and colleagues (Smith et al., 2013) coined the term "co-metabolism" to indicate that foods in humans are being metabolized by both host enzymes and microbial enzymes, and that the products of one are often substrates for the other. We have an entangled metabolism.

Whereas metabolism was originally defined as a cellular phenomenon, metabolism can now be seen as a communal enterprise (Dupré & O'Malley, 2009; Kelty & Landecker, 2019). Living beings are holobionts, organized consortia containing several species. A holobiont is both an organism and a collection of ecological communities (Gilbert, 2019a, 2019b, 2023; Suarez & Stencel, 2020). The symbionts of a holobiont are each other's environment, and symbiotic relations are critical for uniting organisms into their respective environments (Formosinho et al., 2022). About half the cells of the human body are microbial, not derived from the zygote. These microbes are not merely fellow passengers, eating food with us; they are intimately connected with our physiological functioning (McFall-Ngai et al., 2013; Gilbert et al., 2012a, 2012b).

In holobionts, the set of symbiotic organisms (including the "host") participate in each other's metabolisms. This is readily seen in the mealy bug *Planococcus*, where the production of the amino acid phenylalanine begins with enzymes encoded by the bug's symbionts, *Tremblaya* bacteria. The metabolites that are generated by these enzymes then travel into another bacterial species, *Moranella*, which are symbionts within the *Tremblaya* bacteria. The newly made metabolites from *Moranella* then return to the *Tremblaya* bacteria, wherein they are converted into a product that can be metabolized into phenylalanine by enzymes encoded by the *Planococcus* genome (McCutcheon & von Dohlen, 2011).

In mammals, bacterial products, especially the short chain fatty acids produced by the bacterial digestion of plant carbohydrates, are sent through the circulation of the body, where they sustain digestion, blood circulation, and other physiological functions (Kelty & Landecker, 2019; McFall-Ngai et al., 2013). The gut microbes are even critical for producing new neurons in adult mice. Indole, synthesized from dietary tryptophan by the mother's gut microbiome, is carried through the blood to the hippocampus. Here, it activates the aryl hydrocarbon receptor of the neural stem cells, converting the receptor into a functional transcription factor that activates the genes responsible for neuron production (Wei et al., 2021). Since these are the neurons thought to be critical for memory and learning, the bacteria are truly a part of “who” one is. As Wei et al. (2021, p. 2) claim, “Gut microbes are an evolving, prokaryotic component of the meta-organismal self.”

The river of metabolism has many tributaries. Metabolism is co-metabolism. We are all lichens.

5.3 Co-development

The river of development had been seen as a single flow, originating solely in the fertilized egg and cascading into multiple cells types that are arranged into organs. The life cycle was the expected to be the performance of single genome playing many roles: egg, gastrula, ectoderm, mesoderm, ectoderm, germ cells, and finally senescent tissue and death. However, we have recently seen that there are several tributaries that form the river of ontological development. Joining the life cycle of the zygotically derived cells are those cells derived from bacterial and fungal sources. Symbiosis is not only for mutually consenting adults. Development is a team sport, and symbiosis between microbes and multicellular eukaryotic organisms help construct the adults. Co-development is sometimes called “sympoiesis,” in opposition to the notion of autopoiesis (Clarke & Gilbert, 2022; Dempster, 1998; Haraway, 2016). Living entities are not self-constructed. Rather they are created by the interactions of parts that are themselves completed by this process. This notion of reciprocal coming-into-being is a standard feature of vertebrate embryology. The eyes, for instance, originate when cellular bulges from the brain contact the nascent epidermis in two regions of the head. The brain bulges call to the epidermis to become lens tissues. The responding cells, as they differentiate into lens tissues, secrete factors that tell the cells of the brain bulge to become the retina. By continued calls and responses, the lens and the retina co-construct each other (see Barresi & Gilbert, 2023).

Now we find that microbes, including bacteria from another kingdom of life, can act as embryonic cells do and induce gene expression in their neighboring cells. In mice, this was shown by Hooper and Gordon (2001), when they demonstrated that gut bacteria, and particularly *Bacteroides thetaiotamicron*, were capable of inducing the expression of several genes in the Paneth cells of the mouse intestine. One of these genes, the gene encoding Angiogenin-4, was particularly important since this protein is secreted by the gut epithelium and induces the overlying mesoderm to form the gut capillary network (Stappenbeck et al., 2002). Other bacteria (such as *E. coli* and *B. infantis*) can also induce the activation of the Angiogenin-4 gene, but not as efficiently as *B. thetaiotamicron*. Thus, in addition to “It’s The Song, Not the Singer,” (Doolittle & Inkpen, 2018; Papale, this volume) one can add “Nobody Does It Better.”

Together, *Bacteroides thetaiotamicron* and Angiogenin 4 help organize the gut epithelium and the gut microbiome population.

- Angiogenin-4 induces capillary formation, possibly through decay of miR-141 (Weng et al., 2022)

- Angiogenin-4 acts through Wnt and Notch to expand $Lgr5^+$ intestinal stem cells; at higher concentrations it causes apoptosis of these cells, thereby becoming a regulator of intestinal homeostasis (Abo et al., 2023).
- Angiogenin-4 regulates the bacterial population of the gut. It is toxic to *Listeria, Candida* and *Salmonella*, potential competitors of *Bacteroides.* Administration of Ang4 increased beneficial bacteria such as *Lactobacillus and Akkermansia* and also decreased certain pathogenic bacteria, including *Alistipes* and *Enterohabdus* (Sultana et al., 2022). Angiogenin-4 is also toxic to *Bacteroides thetaiotamicron,* thereby limiting its effects.

Host and symbiont work together to structure the intestine and its microbial community, and immunity (the ability of the body to reject certain bacteria while welcoming others) is a property of the holobiont, not merely a property of the host (see Gilbert & Tauber, 2016). Moreover, just as the bacteria can induce changes in the host, so the host induces changes in the bacteria. Physiological concentrations of bile extract from the gall bladder induce the formation of biofilms from *B. thetaiotaomicron* (Béchon et al., 2022).

Whereas numerous species of bacteria appear to be capable of inducing angiogenin-4 in mice, some other inductions appear to be more specific. In zebrafish, the proliferation of insulin-secreting beta-cells in the pancreas is promoted by the BefA protein (Hill et al., 2022). This protein is secreted by only one species of microbes in the water normally associated with zebrafish, *Aeromonas veronii.* Similarly, in the squid *Euprymna,* only one species of bacteria *Vibrio fisheri,* is able to induce the formation of the light organ (Nyholm & McFall-Ngai, 2021). Some species are able to perform different developmental actions in different organisms. *Wolbachia* bacteria are responsible for the proper orientation of the second mitotic division of the nematode *Brugia malayi* (Landman et al., 2014), and they are needed for the formation of ovaries in the *Asobara* wasp (Dedeine et al., 2001). Sympoiesis is found throughout the animal and plant kingdoms (see Gilbert, 2024).

Even mammalian brains are formed through the interaction of many species. Although mammalian embryos are often thought to develop within a sterile amnion (Kennedy et al., 2023), about 30% of the small molecules carried in the maternal bloodstream of a pregnant mouse arise directly or indirectly from symbiotic microbes (McFall-Ngai et al., 2013; Nicholson et al., 2012). In this way, the microbes in the maternal gut can synthesize compounds that circulate throughout the body, cross the placenta, and affect the development of the fetus. When the gut microbes of a pregnant mouse digest plant fiber, they produce short-chain fatty acids (such as butyrate and propionate) that enter the maternal bloodstream, pass into the fetal blood circulation, and reach the organs of her developing embryos. Here, these bacterial products activate particular genes in the embryonic pancreas, nervous system, and intestine. These changes in gene expression generate proteins that mature the sympathetic neurons and confer life-long obesity-resisting metabolic phenotypes to the offspring (Kimura et al., 2020).

Other soluble products of the maternal gut microbiome (especially hippurate and trimethyl-5-aminovalerate) stimulate the maturation of auditory neurons in the fetal mouse brain (Vuong et al., 2020). If these latter metabolites are not produced by the microbiome of the pregnant mouse, the resulting adult progeny have hearing deficiencies. As Sarkis Mazmanian (2024) recently said, "It is remarkable that gut bacteria have figured out how to send signals to the brain from the gut to regulate complex outcomes."

Bacteria have even been seen to allow several marine organisms to complete their life cycle by inducing the settlement and metamorphosis of their larvae. In the life cycle of the sponge *Amphimedon*, for instance, resident symbiotic bacteria supply the arginine that is utilized to synthesize the nitric oxide cue for settlement of the larvae and their metamorphosis into adult sponges (Song et al., 2021). In many species, larvae can't settle and undergo metamorphosis until they find a place where specific bacteria have previously settled. The marine tube worm *Hydroides elegans* will complete its life cycle only after the larvae land on the biofilm-forming bacterium *Cellulophaga lytica*. The lipopolysaccharide (LPS) in the capsule of this bacterium is not only a structural molecule needed for the integrity of the bacterial cell wall; it is also an informational molecule necessary for the settlement and metamorphosis of the larval tube worm (Freckelton et al., 2022). LPS molecules (especially their sugar moieties) differ between different species of bacteria, and the LPS of different bacterial strains appears to induce the settlement and metamorphosis of different invertebrate species (Hu et al., 2024; Freckelton et al., 2022). Thus, larval settlement may depend upon how larvae recognize and interpret the LPS of different bacteria. This, in turn means that the biogeography of many marine organisms (corals, tube worms, etc.) depends on where certain bacteria are located.[1]

The stream of development has many tributaries. Development is co-development. We are adults, and we are also climax communities.

5.4 Co-evolution

Lynn Margulis (1993) made the point that "Life did not take over the globe by combat, but by networking." Nature can produce selectable variation in numerous ways. It can alter the placement, timing, or amount of a gene's expression or even create a new regulatory gene (Arthur, 2011). Or it can generate variation by changing the modules in a holobiont. If evolution in caused by the appearance of selectable changes in development, then such developmental changes can involve the relationships between

[1] Remarkably, the LPS molecule may be a crucial cue for symbiosis even in terrestrial animals. LPS molecule is also used to initiate symbiosis in the *Riptortus* bug (Kim et al., 2016). Here, it is probably used for attachment of the bacterium to the gut wall. Mammals interpret the LPS signal in a critically important manner, where the recognition of LPS usually has been seen to activate the innate immune system. It is also known as entotoxin, an inducer of inflammation. Here, LPS is recognized by and activates the host Toll Receptor-4 protein, initiating the induction several pro-inflammatory, anti-viral, and anti-bacterial cytokines (Soares et al., 2010). Recognition of LPS is critical for keeping gut microbes from spreading elsewhere (Buscher et al., 2016) and is thus a critical control mechanism for regulating symbiont placement.

symbionts and host (the big symbiont). This is readily seen in the microevolutionary contexts, where symbionts can be an important source of selectable variation, especially in parthenogenetic organisms. In pea aphids, for instance, bacterial symbionts can determine the color of the aphid (Tsushida et al., 2010), the thermotolerance of the aphid (Dunbar et al., 2007), and the aphid's ability to resist parasitoid wasp infections (Oliver et al., 2009). In the soy bug *Riptortus*, a stain of *Burkholderia* bacteria that encodes enzymes that digest fenithrothion insecticide provides insecticide resistance for the holobiont containing it (Kikuchi et al., 2012; Lee et al., 2019), and the red turpentine beetle increased its fitness by partnering with a new fungus when it was imported into China (Sun et al., 2013; Taerum et al., 2013).

On the macroevolutionary scale, symbiotic associations of eukaryotic organisms with bacteria has also been postulated to be responsible for the origins of multicellularity (Dayel et al., 2011) and sexual meiosis (Woznica et al., 2017), two of the major transitions of life. There may be other origins of these phenomena; but present-day species of the single-celled choanoflagellates can divide to become multicellular organisms and differentiate haploid gametes when placed into association with these bacteria.

But the organisms that bring together best the three streams of co-metabolism, co-development, and co-evolution are the herds of common cattle, *Bos taurus*. Indeed, for discussions of the origin of herbivory, "*fetchez la vache*!" Herds of had been an important feature of early evolutionary theory, and the ideas of Sewall Wright's, 1931 "Evolution in Mendelian Populations" (small populations, epistasis, shifting balances, and inbreeding coefficients) can be traced back to his earlier (Wright, 1920) volume *Principles of Livestock Breeding*, which he wrote while being Senior Animal Husbandman at the US Department of Agriculture.

First, the metabolic stream is depicted in the symbiosis between cattle and the complex ecosystem of protists, bacteria, and Archaea in their rumens. There are no genes in the bovine genome capable of making the proteins necessary for the digestion or fermentation of plant substances. Those genes and proteins come from the ruminal microbes that hydrolyze complex plant carbohydrates (such as cellulose) into simple sugars, and ferment them into short chain fatty acids such as acetate, butyrate, and propionate. These compounds can be used for aerobic respiration, gluconeogenesis, and lipid formation (Moraïs & Mizrahil, 2019a, 2019b). Cattle are obligate herbivores, and microbes are a crucial and necessary part of the cow. This is nutritional symbiosis, co-metabolism.

Second, the developmental stream involves the way that cows develop. Cows, like most mammals, receive symbionts as they pass through the birth canal. About 1–2 days after birth, microbes can be seen entering the area of the digestive tube that will become the rumen. But these bacteria become metabolically dormant until weaning. As long as the calf is given milk, the milk is shunted to the abomasum portion of the stomach and provides nutrition for the calf. However, upon weaning, the groove transporting the milk disappears and plant material enters into the rumen. Here, the plant material becomes digested and fermented into short chain fatty acids. It is this set of short chain fatty acids–acetate, butyrate, and propionate– that induce the formation of the functional rumen. The rumen grows and develops its characteristic convolutions (Sander et al., 1959; Baldwin & Conner, 2017; Liu et al., 2024). It appears that bacteria such as *Prevotella*, *Bacteroides*, and *Ruminococcus* work together to create short chain fatty acids from cellulose, and these molecules activate the genes for numerous transcription factors that promote the growth and differentiation of rumen tissue (Malmuthuge et al., 2019). Indeed, putting butyrate into the gut will induce rumen formation in calves that have not yet been weaned (Sander et al., 1959).

This new activation of gene expression appears to be accomplished by modifying specific nucleosomes. When butyrate is given to cultured rumen cells in the laboratory, the levels of nucleosome acetylation correspond to the levels of butyrate, and there is a correlation between the levels of butyrate and the developmental maturity of the rumen (Deng et al., 2023; Kang et al., 2023). Thus, the bacteria help construct the rumen, which will house them and allow the survival of the host organism; the microbes build their own home. This sympoiesis is niche construction within the organism. This is sympoiesis, co-development.

Last, this developmental change, producing the rumen, has promoted the evolution of a major new clade of animals–the ruminants. Ruminants have been remarkably successful terrestrial herbivores, and they have diversified into cattle, buffalo, yak, sheep, goats, pronghorns, giraffes, antelopes, okapis, deer, and chevrotains. These nutritional and developmental symbioses mentioned above have created a new dietary niche for these ruminants through two types of niche construction, as recognized by Lewontin (1982). First, there is the above-mentioned "perturbational niche construction," whereby gut microbes initiate the building of a mature rumen from the non-functional anlagen of this stomach region. Second, there is "mediational niche construction," whereby microbe-induced changes alter how the animal experiences environmental resources without actual modification of the environment. Thanks to the mutualistic microbes, plants are now edible. Such reciprocal niche construction between the host and its associated microbial organisms scaffold each other's developmental and phenotypic processes as well as organize a new selective environment of the holobiont (Chiu & Gilbert, 2020). As Vermeij and Lindburg (2000) note, "The establishment of symbioses between potential herbivores and various microorganisms in the gut may have been a necessary precondition for the evolution of feeding on such plant tissues." The genes constructing the rumen are mostly older genes recruited for a new function (Chen et al., 2022). It is thought that the rumen originated 35 million years ago from distal portions of the esophagus and became contiguous with the stomach.

Moreover, the population of microbes in the rumen can change rapidly and adopt the organism to new diets as seasons and locales change (Fernando et al., 2010; Sasson et al., 2022). Kauffman and Johnson (1991) wrote that "Coevolution is thought of, at the lowest level, as a coupling of landscapes such that adaptive moves by one player deform the landscapes of its immediate partners." In the case of cows, that landscape is not merely metaphorical. It is quite literal. This is evolution by niche construction within and outside the organism.

There are rivers within rivers. Each organism finds itself simultaneously within the metabolic river of daily life, the developmental river of ontogenesis, and the evolutionary river reaching back billions of years. And no river has a single source. We metabolize with partners; we develop with partners, and we evolve with partners. As Longfellow (1842) wrote, "Thou hast taught me, Silent River, many a lesson, deep and long." And thank you, John, for being one of the great expositors, one of the great interpreters, of these rivers.

Acknowledgements I wish to thank Sabina Leonelli for the invitation to write for this event.

References

Abo, H., Sultana, M. F., & Kawashima, H. (2023). Dual function of angiogenin-4 inducing intestinal stem cells and apoptosis. *Frontiers Cell and Developmental Biology, 11*, 1181145. https://doi.org/10.3389/fcell.2023.1181145

Arthur, W. (2011). *Evolution: A developmental approach.* Wiley-Blackwell.

Baldwin, R. L. IV., & Conner, E. E. (2017). Rumen function and development. *Veterinary Clinics: Food Animal Practice, 33*, 427–439. https://doi.org/10.1016/j.cvfa.2017.06.001

Barresi, M. J. F., & Gilbert, S. F. (2023). *Developmental biology.* Oxford University Press.

Béchon, N., Mihajlovic, J., Lopes, A. A., Vendrell-Fernández, S., Deschamps, J., et al. (2022). *Bacteroides thetaiotaomicron* uses a widespread extracellular DNase to promote bile-dependent biofilm formation. *Proceedings. National Academy Science USA, 119*(7), Article e2111228119. https://doi.org/10.1073/pnas.2111228119

Bernard, C. (1974). *Lessons on the phenomena of life common to animals and plants.* Translated by R. E. Hoff, R. Guillemin, and L. Guillemin. Thomas.

Buscher, K., Wang, H., Zhang, X., Striewski, P., Wirth, B., et al. (2016). Protection from septic peritonitis by rapid neutrophil recruitment through omental high endothelial venules. *Nature Communications, 7*, 10828.

Chen, C., Yin, Y., Li, H., Zhou, B., Zhou, J., Zhou, X., Li, Z., et al. (2022). Ruminant-specific genes identified using high-quality genome data and their roles in rumen evolution. *Science Bulletin (Beijing), 67*(8), 825–835. https://doi.org/10.1016/j.scib.2022.01.02

Chiu, L., & Gilbert, S. F. (2020). Niche construction and the transition to herbivory: Phenotype switching and the origination of new nutritional modes. In H. Levine, M. Jolly, P. Kulkarni, & V. Nanjundiah (Eds.), *Phenotype switching: Implications in biology and medicine* (pp. 459–482). Elsevier.

Clarke, B., & Gilbert, S. F. (2022). Margulis, autopoiesis, and sympoiesis. In C. A. Jones, N. Bell, & S. Nimrod (Eds.), *Symbionts: Contemporary artists and the biosphere* (pp. 63–78). MIT Press.

Dayel, M. J., Alegado, R. A., Fairclough, S. R., Levin, T. C., Nichols, S. A., McDonald, K., & King, N. (2011). Cell differentiation and morphogenesis in the colony-forming choanoflagellate *Salpingoeca rosetta. Developmental Biology, 357*, 73–82.

Dedeine, F., Vavre, F., Fleury, F., Loppin, B., Hochberg, M. E., & Bouletreau, M. (2001). Removing symbiotic *Wolbachia* bacteria specifically inhibits oogenesis in a parasitic wasp. *Proceedings. National Academy Science USA, 98*, 6247–6252. https://doi.org/10.1073/pnas.101304298

Dempster, M. B. (1998). *A self-organizing systems perspective on planning for sustainability.* MA thesis, Environmental Studies, University of Waterloo, 1998. http://www.bethd.ca/pubs/mesthe.pdf. Cited in Haraway, *op.cit.* p. M27.

Deng, J., Liu, Y. J., Wei, W. T., Huang, Q. X., Zhao, L. P., Luo, L. Y., et al. (2023). Single-cell transcriptome and metagenome profiling reveals the genetic basis of rumen functions and convergent developmental patterns in ruminants. *Genome Research, 33*, 1690–1707. https://doi.org/10.1101/gr.278239.123

Doolittle, W. F., & Inkpen, S. A. (2018). Processes and patterns of interaction as units of selection: An introduction to ITSNTS thinking. *Proceedings. National Academy Science USA, 115*, 4006–4014. https://doi.org/10.1073/pnas.1722232115

Dunbar, H. E., Wilson, A. C. C., Ferguson, N. R., & Moran, N. A. (2007). Aphid thermal tolerance is governed by a point mutation in bacterial symbionts. *PLoS Biology, 5*(5), Article e96.

Dupré, J., & O'Malley, M. (2009). Varieties of Living things: Life at the Intersection of lineage and metabolism. *Philosophy and Theory in Biology, 1*, Article e0003.

Fernando, S. C., Purvis, H. T., Najar, F. Z., Sukharnikov, L. O., Krehbiel, C. R., et al. (2010). Rumen microbial population dynamics during adaptation to a high-grain diet. *Applied and Environmental Microbiology, 76*, 7482–7490. https://doi.org/10.1128/AEM.00388-10

Florkin, M., & Stotz, E. M. (1973). Crystallization and biosynthesis. *Comparative Biochemistry, 32*, 133–143.

Formosinho, J., Bencard, A., & Whiteley, L. (2022). Environmentality in biomedicine: Microbiome research and the perspectival body. *Studies in History of Philosophy of Science, 91*, 148–158. https://doi.org/10.1016/j.shpsa.2021.11.005

Freckelton, M. L., Nedved, B. T., Cai, Y.-S., Cao, S., Turano, H., Alegado, R. A., & Hadfield, M. G. (2022). Bacterial lipopolysaccharide induces settlement and metamorphosis in a marine larva. *Proceedings of the National Academy of Sciences USA, 119*, Article e2207951119. https://doi.org/10.1073/pnas.2200795119

Gilbert, S. F. (1982). Intellectual traditions in the life sciences: Molecular biology and biochemistry. *Perspectives in Biology and Medicine, 26*, 151–162.

Gilbert, S. F. (2019a). Towards a developmental biology of holobionts. In G. Fusco (Ed.), *Perspectives on evolutionary and developmental biology: Essays for Alessandro Minelli* (pp. 13–22). Padova University Press.

Gilbert, S. F. (2019b). Evolutionary transitions revisited: Holobiont evo-devo. *Journal of Experimental Zoology Molecular and Developmental Biology, 332*, 307–314. https://doi.org/10.1002/jez.b.22903

Gilbert, S. F. (2023). Inter-kingdom communication and the sympoietic way of life. *Frontiers in Cell and Developmental Biology, 12*, 1427798. https://doi.org/10.3389/fcell.2024.1427798

Gilbert, S. F. (2024). Holobiont development: Embryology and ecological succession. *Human Development, 67*, 257–272. https://doi.org/10.1159/000534203

Gilbert, S. F., Sapp, J., & Tauber, A. I. (2012a). A symbiotic view of life: We have never been individuals. *Quarterly Review of Biology, 87*, 325–341. https://doi.org/10.1086/668166

Gilbert, S. F., Sapp, J., & Tauber, A. I. (2012b). A symbiotic view of life: We have never been individuals. *Quarterly Review of Biology, 87*, 325–341.

Gilbert, S. F., & Tauber, A. I. (2016). Rethinking individuality: The dialectics of the holobiont. *Biology and Philosophy, 31*, 839–853.

Guggenheim, K. Y. (1991). Rudolf Schoenheimer and the concept of the dynamic state of body constituents. *Journal of Nutrition, 121*(11), 1701–1704.

Haldane, J. B. S. (1932). *The origin of life*. In *The Inequality of man and other essays*. Chatte & Windus. https://www.uv.es/~orilife/textos/Haldane.pdf

Haraway, D. J. (2016). *Staying with the trouble: Making Kin in the Chthulucene*. Duke University Press.

Hill, J. H., Massaquoi, M. S., Sweeney, E. G., Wall, E. S., Jahl, P., Bell, R., et al. (2022). BefA, a microbiota-secreted membrane disrupter, disseminates to the pancreas and increases β cell mass. *Cell Metabolism, 34*, 1779-1791.e9. https://doi.org/10.1016/j.cmet.2022.09.001

Holmes, F. L. (1991). *Hans Krebs: The Formation of a scientific life, 1900–1933* (Vol. 1). Oxford University Press.

Hooper, L. V., & Gordon, J. I. (2001). Commensal host-bacterial relationships in the gut. *Science, 292*(5519), 1115–1118. https://doi.org/10.1126/science.1058709

Hu, X. M., Peng, L., Wu, J., Wu, G., Liang, X., & Yang, J. L. (2024). Bacterial c-di-GMP signaling gene affects mussel larval metamorphosis through outer membrane vesicles and lipopolysaccharides. *NPJ Biofilms Microbiomes, 10*, 38. https://doi.org/10.1038/s41522-024-00508-6

Huxley, T. H. (1870). Address to the British Association: Liverpool Meeting, 1870. *Nature, 2*, 400–406.

Jonas, H. (1966). *The phenomenon of life*. Harper and Row.

Kang, X., Li, C., Liu, S., Baldwin, R. L., Liu, G. E., & Li, C. J. (2023). Genome-Wide acetylation modification of H3K27ac in bovine rumen cell following butyrate exposure. *Biomolecules, 13*(7), 1137. https://doi.org/10.3390/biom13071137

Kauffman, S. A., & Johnson, S. (1991). Coevolution to the edge of chaos: Coupled fitness landscapes, poised states, and co-evolutionary avalanches. *Journal of Theoretical Biology, 149*, 467–505.

Kelty, C., & Landecker, H. (2019). Outside In: Microbiomes, epigenomes, visceral sensing, and metabolic ethics. In *After practice: Thinking through matter(s) and meaning relationally,* The Laboratory for the Anthropology of the Environment and Human Relations (pp. 53–65). Panama Verlag.

Kennedy, E. P. (2001). Hitler's gift and the era of biosynthesis. *Journal of Biological Chemistry, 276*, 42619–42631.

Kennedy, K. M., de Goffau, M. C., Perez-Muñoz, M. E., Arrieta, M. C., Bäckhed, F., et al. (2023). Questioning the fetal microbiome illustrates pitfalls of low-biomass microbial studies. *Nature, 613*, 639–649. https://doi.org/10.1038/s41586-022-05546-8

Kikuchi, Y., Hayatsu, M., Hosokawa, T., Nagayama, A., Tago, K., & Fukatsu, T. (2012). Symbiont-mediated insecticide resistance. *Proceedings National Academy of Sciences USA, 109*, 8618–8623.

Kim, J. K., Park, H. Y., & Lee, B. L. (2016). The symbiotic role of O-antigen of *Burkholderia* symbiont in association with host *Riptortus pedestris. Developmental and Comparative Immunology, 60*, 202–208. https://doi.org/10.1016/j.dci.2016.02.009

Kimura, I., Miyamoto, J., Ohue-Kitano, R., Watanabe, K., Yamada, T., Onuki, M., Aoki, R., et al. (2020). Maternal gut microbiota in pregnancy influences offspring metabolic phenotype in mice. *Science, 367*, eaaw8429. https://doi.org/10.1126/science.aaw8429

Landecker, H. (2013). The metabolism of philosophy, in three parts. In B. Malkmus & I. Cooper (Eds.), *Dialectic and paradox: Configurations of the third in modernity* (pp. 193–224). Peter Lang.

Landecker, H. (2017). Metabolism, autonomy, and individuality. In S. Gissis, E. Lamm, & A. Shavit (Eds.), *Landscapes of collectivity in the life sciences* (pp. 225–248). MIT Press.

Landmann, F., Foster, J. M., Michalski, M. L., Slatko, B. E., Sullivan, W., & W. (2014). Co-evolution between a nematode and its nematode host: *Wolbachia* asymmetric localization and A-P polarity establishment. *PLoS Neglected Tropical Diseases, 8*(8), Article e3096.

Lee, J., Kim, C. H., Jang, H. A., Kim, J. K., Kotaki, T., Shinoda, T., Shinada, T., Yoo, J. W., & Lee, B. L. (2019). *Burkholderia* gut symbiont modulates titer of specific juvenile hormone in the bean bug *Riptortus pedestris. Developmental and Comparative Immunology, 99*, Article 103399. https://doi.org/10.1016/j.dci.2019.103399

Lewontin, R. (1982). Organism and environment. In H. C. Plotkin (Ed.), *Learning, development and culture* (pp. 151–170). Wiley.

Liu, W., Yan, H., Jia, W., Huang, J., Fu, Z., Xu, Z. W., et al. (2024). Association between gut microbiota and Hirschsprung disease: A bidirectional two-sample Mendelian randomization study. *Frontiers of Microbiology, 15*, 1366181. https://doi.org/10.3389/fmicb.2024.1366181

Longfellow, H. W. (1842). To the River Charles. In *Ballads and other Poems*. John Owen. https://www.hwlongfellow.org/poems_poem.php?pid=74

Malmuthuge, N., Liang, G., & Guan, L. L. (2019). Regulation of rumen development in neonatal ruminants through microbial metagenomes and host transcriptomes. *Genome Biology, 20*, 172. https://doi.org/10.1186/s13059-019-1786-0

Margulis, L. (1993). *Microbial microcosm*. https://www.context.org/iclib/ic34/margulis/

Mazmanian, S. (2024). Latest research on the gut-brain axis. *Seed*. https://seed.com/cultured/latest-research-on-the-gut-brain-axis/

McCutcheon, J. P., & von Dohlen, C. D. (2011). An interdependent metabolic patchwork in the nested symbiosis of mealybugs. *Current Biology, 21*, 1366–1372.

McFall-Ngai, M., Hadfield, M. G., Bosch, T. C. G., Carey, H. V., Domazet-Lošo, T., Douglas, A. E., et al. (2013). Animals in a bacterial world, a new imperative for the life sciences. *Proceedings National Academy Science USA, 110*(9), 3229–3236. https://doi.org/10.1073/pnas.1218525110

Meincke, A. S. (2022). Biological subjectivity: Processual animalism as a unified account of personal identity. In J. Noller (Ed.), *The unity of a person: Philosophical perspectives* (pp. 100–126). Routledge.

Mendelsohn, E. (1965). Physical models and physiological concepts: Explanation in nineteenth century biology. *British Journal for the History of Science, 2*, 201–218.

Monod, J. (1971). *Chance and necessity*. Knopf.

Moraïs, S., & Mizrahi, I. (2019a). Islands in the stream: from individual to communal fiber degradation in the rumen ecosystem. *FEMS Microbiology Reviews*. https://doi.org/10.1093/femsre/fuz007

Moraïs, S., & Mizrahi, I. (2019b). The road not taken: The rumen microbiome, functional groups, and community states. *Trends in Microbiology, 27*, 538–549. https://doi.org/10.1016/j.tim.2018.12.011

Nicholson, D. J. (2018). Reconceptualizing the organism. In D. J. Nicholson & J. Dupre (Eds.), *Everything flows: Towards a processual philosophy of biology* (pp. 139–166). Oxford University Press.

Nicholson, D. J., & Dupre, J. (2018). *Everything flows: Towards a processual philosophy of biology*. Oxford University Press.

Nicholson, J. K., Holmes, E., Kinross, J., Burcelin, R., Gibson, G., Jia, W., & Pettersson, S. (2012). Host-gut microbiota metabolic interactions. *Science, 336*(6086), 1262–1267. https://doi.org/10.1126/science.1223813

Nyholm, S. V., & McFall-Ngai, M. (2021). A lasting symbiosis: How the Hawaiian bobtail squid finds and keeps its bioluminescent bacterial partner. *Nature Reviews Microbiology, 19*, 666–679. https://doi.org/10.1038/s41579-021-00567-y

Olby, R. (1971). *The path to the double helix*. Macmillan.

Oliver, K. M., Degnan, P. H., Hunter, M. S., & Moran, N. A. (2009). Bacteriophages encode factors required for protection in a symbiotic mutualism. *Science, 325*, 992–994.

Pfeiffer, J. E. (1960). DNA. *Natural History, 69*, 8–15.

Ritterbush, P. C. (1968). *The art of organic forms*. Smithsonian Institution Press.

Sander, E. G., Warner, R. G., Harrison, H. N., & Loosli, J. K. (1959). The stimulatory effect of sodium butyrate and sodium propionate on the development of rumen mucosa in the young calf. *Journal of Dairy Science, 42*, 1600–1605. https://doi.org/10.3168/jds.s0022-0302(59)90772-6

Sasson, G., Moraïs, S., Kokou, F., Plate, K., Trautwein-Schult, A., Jami, E., Bayer, E. A., Becher, D., & Mizrahi, I. (2022). Metaproteome plasticity sheds light on the ecology of the rumen microbiome and its connection to host traits. *ISME Journal, 16*(11), 2610–2621. https://doi.org/10.1038/s41396-022-01295-8

Schoenheimer, R. (1942). *The dynamic state of body constituents*. Harvard University Press.

Schoenheimer, R., & Rittenberg, D. (1935). Deuterium as an indicator in the study of intermediary metabolism. *Science, 82*(2120), 156–157. https://doi.org/10.1126/science.82.2120.156

Schwann, T. (1839). Mikroskopick Untersuchungen über die Uebereinstimmung in der Struktur und dem Wachstehun der Thiere und Pflanzen.

Smith, M. I., Yatsunenko, T., Manary, M. J., Trehan, I., Mkakosya, R., et al. (2013). Gut microbiomes of Malawian twin pairs discordant for Kwashiorkor. *Science, 339*(6119), 548–554. https://doi.org/10.1126/science.1229000

Soares, J. B., Pimentel-Nunes, P., Roncon-Albuquerque, R., & Leite-Moreira, A. (2010). The role of lipopolysaccharide/toll-like receptor 4 signaling in chronic liver diseases. *Hepatology International, 4*(4), 659–672. https://doi.org/10.1007/s12072-010-9219-x

Song, H., Hewitt, O. H., & Degnan, S. M. (2021). Arginine biosynthesis by a bacterial symbiont enables nitric oxide production and facilitates larval settlement in the marine-sponge host. *Current Biology, 31*, 433–437. https://doi.org/10.1016/j.cub.2020.10.051

Stappenbeck, T. S., Hooper, L. V., & Gordon, J. I. (2002). Developmental regulation of intestinal angiogenesis by indigenous microbes via Paneth cells. *Proceedings. National Academy Science USA, 99*, 15451–15455. https://doi.org/10.1073/pnas.202604299

Suarez, J., & Stencel, A. (2020). A part-dependent account of biologic individuality: Why holobionts are individuals and ecosystems simultaneously. *Biology Reviews, 95*, 1308–1324. https://doi.org/10.1111/brv.12610

Sultana, M. F., Suzuki, M., Yamasaki, F., Kubota, W., Takahashi, K., Abo, H., & Kawashima, H. (2022). Identification of crucial amino acid residues for antimicrobial activity of angiogenin 4 and its modulation of gut microbiota in mice. *Frontiers in Microbiology, 13*, Article 900948. https://doi.org/10.3389/fmicb.2022.900948

Sun, J., Lu, M., Gillette, N. E., & Wingfield, M. J. (2013). Red turpentine beetle: Innocuous native becomes invasive tree killer in China. *Annual Reviews of Entomology, 58*(1), 293–311. https://doi.org/10.1146/annurev-ento-120811-153624

Taerum, S. J., Duong, T. A., de Beer, Z., Gillette, N., Sun, J.-H., Owen, D. R., & Wingfield, M. J. (2013). Large shift in symbiont assemblage in the invasive red turpentine beetle. *PLoS ONE, 8*(10), Article e78126.

Tsuchida, T., Koga, R., Horikawa, M., Tsunoda, T., Maoka, T., et al. (2010). Symbiotic bacterium modifies aphid body color. *Science, 330*, 1102–1104.

Vermeij, G. J., & Lindberg, D. R. (2000). Delayed herbivory and the assembly of marine benthic ecosystems. *Paleobiology, 26*, 419–430. https://doi.org/10.1666/0094-8373(2000)026%3c0419:DHATAO%3e2.0.CO;2

Vuong, H. E., Pronovost, G. N., Williams, D. W., Coley, E. J. L., Siegler, E. L., et al. (2020). The maternal microbiome modulates fetal neurodevelopment in mice. *Nature, 586*, 281–286. https://doi.org/10.1038/s41586-020-2745-3

Watson, J. D., & Crick, F. H. C. (1953). Molecular structure of nucleic acids: A structure for deoxyribose nucleic acid. *Nature, 171*, 737–738. https://doi.org/10.1038/171737a0

Watson, J. D., & Crick, F. H. C. (1953). The structure of DNA. *Cold Spring Harbor Symposium on Quantitative Biology, 18*, 123–131.

Wei, G. Z., Martin, K. A., Xing, P. Y., Agrawal, R., Whiley, L., et al. (2021). Tryptophan-metabolizing gut microbes regulate adult neurogenesis via the aryl hydrocarbon receptor. *Proceedings of the National Academy of Sciences of the United States of America, 118*(27), Article e2021091118. https://doi.org/10.1073/pnas.2021091118

Weng, C., Dong, H., Bai, R., Sheng, J., Chen, G., Ding, K., Lin, W., Chen, J., & Xu, Z. (2022). Angiogenin promotes angiogenesis via the endonucleolytic decay of miR-141 in colorectal cancer. *Molecular Therapy Nucleic Acids, 27*, 1010–1022. https://doi.org/10.1016/j.omtn.2022.01.017

White, A. P., Handler, & Smith, E. L. (1959). *Biochemistry*. McGraw-Hill.

Woznica, A., Gerdt, J. P., Hulett, R. E., Clardy, J., & King, N. (2017). Mating in the closest living relatives of animals is induced by a bacterial chondroitinase. *Cell, 170*, 1175-1183.e11.

Wright, S. (1920). *Principles of livestock breeding*. US Department of Agriculture.

Wright, S. (1931). Evolution in mendelian populations. *Genetics, 16*(2), 97–159. https://doi.org/10.1093/genetics/16.2.97

Yoxen, E. J. (1979). Where does Schroedinger's 'What is Life' belong in the history of molecular biology? *History of Science, 17*(17–52), 1979.

Chapter 6
Is ITSNTS Just an Instance of Process Epistemology?

François Papale, W. Ford Doolittle, and S. Andrew Inkpen

Abstract We argue that there is an explanatory or epistemic value in assuming that "It's the song not the singer(s)" or ITSNTS is an example of process, rather than thing or substance, epistemology and that process ontology as in Nicholson and Dupré is the preferred way to go. We see this as one way to defeat the Platonic claim that ITSNTS is immaterial but affects, causally, the world of "things".

6.1 Introduction

We argue that the evolution of multispecies consortia is better explained as the evolution of processes, rather than of things or substances. This was the point of the "It's the song, not the singer(s)" (ITSNTS) theory that two of us elaborated in 2018 (Doolittle & Inkpen, 2018), and which originated in 2010 (Doolittle & Zhaxybayeva, 2010). And if we're correct, this reveals an *explanatory* or *epistemic* benefit of a process philosophy for evolutionary biology. In this chapter, we'd like to go some way towards that revelation by explaining our thinking and defending it against a common objection to ITSNTS.

We should make clear at the outset, though, that none of us is a metaphysician and our concerns here are with process philosophy as an *epistemic* (explanatory) project. We're not concerned with the question of whether the world is made up of

F. Papale
Department of Philosophy, Université McGill, Montréal, Canada
e-mail: francoisppl@gmail.com

W. F. Doolittle (✉)
Department of Biochemistry and Molecular Biology, Dalhousie University, Halifax, NS, Canada
e-mail: W.Ford.Doolittle@Dal.Ca

Department of Philosophy, Dalhousie University, Halifax, NS, Canada

S. A. Inkpen
Department of Philosophy, Mount Allison University, Sackville, NB, Canada
e-mail: ainkpen@mta.ca

S. Leonelli et al. (eds.), *Life Beyond Human Boundaries: John Dupré's Philosophy and the Interdisciplinary Study of the Life Sciences*,
https://doi.org/10.1007/978-3-032-26977-5_6

only processes or only things, nor with which is ontologically prior. Like Brunet (2024), we can see how both positions could be fruitful for understanding different living phenomena. We are therefore concerned only with showing how a process-philosophy perspective on *some* evolutionary and ecological phenomena is fruitful, epistemically.

Given that the content of process philosophy varies among its proponents (Dupré & Nicholson, 2018), we need to be clear about how we use these terms, which we take to be mostly in line with commonsense and common parlance in either discipline. Processes, as we conceive them, are characterized by constant change and have temporal parts—breathing, growing, cycling and evolving. Things, or substances, are characterized by stability and durability, or are studied as such. Adult animals, chairs, houses, and mountains have traditionally been understood as static entities whose properties may change. We won't worry here if what most of us unreflectively thought were things—such as organisms or ecosystems—are in fact processes, ontologically. Following Simons (2018), we will say that things are *precipitates* of processes, since this is closest to how we conceived of ITSNTS theory. Simons writes,

> … [things] have causal powers, even though they are not themselves causes. The stone against which I stub a toe cannot be faulted for concreteness; it is precisely because the event of my toe's colliding with it causes pain that I am acutely aware of it as a very solid physical body. For that reason, I prefer to say that [things] are not abstracted from processes but are rather *precipitates* of processes: they are what abides, as certain kinds of processes continue and develop. (Simons, 2018, p. 55)

This is, no doubt, much too little metaphysics for processual philosophers. But, it seems to us that most scientists have tended towards substance metaphysics (Dupré, 2017), and most evolutionists in particular conceive of evolution as something that happens to *things*. In this context, even a limited change of perspective can go a long way, we think. Perhaps some entities will always be described as thing-like, but sometimes process-like explanations have greater value, epistemically.

To sustain our claim that multispecies consortia are best theorized as processes, following the ITSNTS theory, this chapter is structured as follows. In Sects. 2 and 3, we argue against the hologenome theory and how it has initially attempted to embed holobionts, which are multispecies consortia, into evolutionary thinking. In Sect. 2, we suggest that the claim that they are units of selection conflates ecology with evolutionary biology. In Sect. 3, we reiterate that holobionts *qua* things have "too many parents" to be units of selection as classically defined (Godfrey-Smith, 2009). In Sects. 4 and 5, we present an objection that is often voiced against ITSNTS, i.e., that it implies impossible interactions between material and immaterial entities. In Sect. 4, we review the core principles of ITSNTS and in Sect. 5 we tie them to our epistemically-minded and thoroughly material understanding of process ontology, thereby answering the objection. Finally, in Sect. 6, we argue that the interactor/replicator framework, which is compatible with ITSNTS, better fits in a process-based framework for ENS than does the classical Lewontin Recipe. This ties the chapter together by showing how holobionts, when conceived as songs/processes, can be construed in an evolutionarily relevant way that fits already available views of evolution by natural selection.

6.2 Many Holobionts, as Things Made of Things, Cannot Be Units of Selection As Classically Defined

Until recently, change in the structure or function of multispecies consortia—groups of organisms from different species living together—was viewed as a mere byproduct of past selection acting on that community's parts. According to this "classical" view, natural selection acts on populations of individual organisms of the same species only, and identifiable ecosystem or community-level functions bear, at best, the *fortuitous benefits* of this lower-level selection (Godfrey-Smith, 2015).

The hologenome theory of evolution was meant, according to some proponents, to challenge this idea and, given its popularity, was to some extent successful. Holobionts are multispecies consortia, hosts plus their associated microbes. The human microbiome plus the human that harbors it is perhaps the premier example. The central claims of the hologenome theory, to quote its originators, Eugene Rosenberg and Ilana-Zilber Rosenberg, was that:

> The hologenome is defined as the sum of the genetic information of the host and its microbiota. In the hologenome theory of evolution, we suggest that the holobiont (the host and its symbiotic microbiota) with its hologenome, acting in consortium, should be considered a *unit of selection* in evolution, and that relatively rapid variation in the diverse microbial symbionts can have an important role in the adaptation and evolution of the holobiont. (Zilber-Rosenberg & Rosenberg, 2008, p. 723, emphasis ours)

According to what we think to be the predominant "classical" understanding of evolution by natural selection (ENS), this hologenome theory is, however, deeply problematic (Godfrey-Smith, 2009). For all its romantic appeal and the current popularity of the concept underlying it, it is surely wrong to simply state that holobionts can be "units of selection" (Madhusoodanan, 2019; Doolittle & Booth, 2017; Doolittle & Inkpen, 2018; Doolittle, 2024). This is because ENS, as classically understood, requires vertical transmission, or at least some restriction on the independent evolution of involved lineages. In many cases, symbionts must be passed directly from parent hosts to offspring hosts. In other words, the consortium itself must be a part of a parent-offspring lineage to satisfy heredity, which is considered to be a necessary condition of ENS (Lewontin, 1970). But paradigm holobionts, such as the squid-*Vibrio* symbiont or the human microbiome, involve horizontal transmission: they are assembled anew each generation from an environmental pool containing both hosts and symbionts, as independent lineages under separate selection (Doolittle & Booth, 2017).

One problem, which has been highlighted by critics of the hologenome theory, is that the traits and formation of consortia might in fact answer to ecological principles, such as assembly rules and lottery effects, but they do not answer to evolutionary principles. For instance, Queller and Strassmann simply conclude that "complex holobionts can be viewed as ecological communities" (Queller & Strassman, 2016, p. 867). Douglas and Werren suggest that "the concepts and methods of ecology, genetics and evolutionary biology will continue to provide a well-grounded intellectual framework for researching host-microbiome communities, without recourse to the limiting assumption that selection operates predominantly at the holobiont level" (Douglas & Werren, 2016, 10.1128/mbio.02099-15). And Moran and Strauss note with perhaps more nuance that "coevolution and codiversification are likely common for hosts and associated microorganisms. But the proposals that the hologenome is the primary unit of selection and that microbes can be equated to genes within a larger genome are far more sweeping claims" (Moran & Sloan, 2015, e1002311). The target of these critiques is always that holobionts, including the most interesting cases, are *transmitted horizontally*.

So, according to its critics, the hologenome theory, possibly unknowingly, conflates ordinary standard procedures in *ecology* with common thinking in *evolutionary biology* (Skillings, 2016). According to the former discipline, it is the species (prokaryotes or their hosts) which are "assembled", while according to the latter discipline it is alleles within species that *are selected for* (Götzenberger et al., 2012; Morris, 2018; Linquist, 2019; Doolittle, 2024).

We believe that there is a way out of this conflation, but this requires moving beyond the thing-philosophy that prevails in evolutionary biology (Dupré and Nicholson, 2018). The usual way to conceive of multispecies consortia holds on to a thing ontology and epistemology. Indeed, holobionts are often described as static units-of-selection-like entities. They are also described in terms of enduring constitutive parts, such as genes or taxa. This last point leads to the problem described in more detail in the next section.

In contrast, by adopting the ITSNTS perspective and its inherent processual epistemology, *patterns of interaction* between species in a consortium, rather than the species themselves, can then be treated as the relevant *units of selection.* Until proponents of the hologenome theory adopt this essentially ecological perspective, we believe that their thing-based approach to the evolution of populations of holobionts is susceptible to various objections, including the very important one according to which holobionts have "too many parents" (see next section).

6.3 A Last-Ditch (and Failed) Effort to Save Holobionts as Things (Epistemically)

The philosopher Elliott Sober has suggested (personal communication) that holobionts and other ecosystems are just like sexual species, only with more parents. If holobionts can be treated in an evolutionarily-sound perspective as sexual organisms are, then surely they can be conceived as *things* without conflating evolution and ecology. Sober's argument, as we understand it, goes as follows:

> In asexual species, there is only one parent. In sexual species there are two, and in sexual animals these two are generally the sperm donor and the egg donor. Recombination means that there are usually four grand-parents, eight great-grand parents and so on, but in terms of parents, there are only two (ignoring holobiosis for now). Sperms and eggs may have some independent life, sometimes developing into whole (haploid) organisms and often undergoing selection as such. But in the end there are only two parents. In horizontally transmitted holobionts, as in other ecosystems, there are often many more than two– and sometimes hundreds of–"parents". Every lineage that contributes to the formation of a holobiont or ecosystem is the equivalent of a gamete in a biparental sexual organism.

This is where Sober and the thing-ontology approach to multispecies consortia run into a problem. There are "too many parents". Evolution by natural selection is classically presented as a three part recipe, known as Lewontin's Recipe, involving variation, heredity, and fitness difference. This increasingly fails as horizontal transfer and the number of parents increases (Doolittle & Inkpen, 2018). Heritability fails, since offspring consortia become less and less similar to any *particular* parental lineage. Also variability likely fails, since consortia become more and more like each other and to the average of the total of all lineages, if there are too many of these "parents" (analogously to and more extreme than blending inheritance). Selection on any particular lineages' trait weakens. This is just a matter of size of sample (the number of lineages). It's the "law of large numbers" (Doolittle & Inkpen, 2018; Doolittle, 2024).

So unless some principle, equivalent to an "ecological assembly rule" and emphasizing the trait-specific functionality of the lineage intervenes, there will be "too many parents". But this makes the phenomenon leading to a specific holobiont trait an ecological one, not evolutionary, hence the conflation referred to previously. The philosopher of natural selection Peter Godfrey-Smith says of this situation:

> … there must be some parent – offspring similarity, and the clarity of a 'parent-offspring' relation of the relevant kind is inversely related to the number of parents – if there are too many parents there are no parents at all. (Godfrey-Smith 2012, p. 2164)

So this does not work, as far as the classical form of Lewontin's Recipe goes. What we wish to show, however, is that this is the result of focusing on holobionts and their parts as things. If we focus on the processes from which they are precipitated (to reuse Simon's expression) instead, then we can abstract away the transmission patterns of the components and reconcile the ecology of multispecies consortia with evolutionary theory in a way that avoids simple conflation.

6.4 ITSNTS and the Common Objection

"It's the song, not the singer(s)" (ITSNTS) theory aimed to *explain* the evolution of multispecies consortia, and was meant to serve as an alternative to the "hologenome theory of evolution". ITSNTS theory also attempted to unite evolutionary biology and ecology, but more transparently and deliberately than the hologenome theory (Doolittle & Zhaxybayeva, 2010).

It was perhaps *functional redundancy*, i.e. the potential substitution of one functionally equivalent taxon for another, that led to the formulation of ITSNTS in the first place (Doolittle & Zhaxybayeva, 2010; Doolittle & Booth, 2017; Doolittle & Inkpen, 2018). Early on in the development of metagenomics, Turnbaugh and colleagues wrote:

> … it appears that a core gut microbiome exists at the level of shared genes, including an important component involved in various metabolic functions. This conservation suggests a high degree of redundancy in the gut microbiome and supports an ecological view of each individual as an "island" inhabited by unique collections of microbial phylotypes: as in actual islands, different species assemblages converge on shared core functions provided by distinctive components. (Turnbaugh et al., 2009, pp. 483–484)

Closely-related species can often substitute for each other because they share the same relevant function, and genes encoding a relevant function can also be transferred laterally between unrelated species: there *is* taxonomic redundancy, and—at a certain level—functional constancy (Tian et al., 2020).

This functional redundancy runs counter to what we would expect for holobionts if we needed to keep a thing ontology and epistemology to study them. Substance philosophy focuses on *things*, such as genes and organisms. And any particular story of reproduction and subsequent lineage formation is very much about things. The question regarding the evolutionary stability of holobionts would be the presence or absence of this or that gene, of this or that taxon. But both ITSNTS and cutting-edge research regarding multispecies consortia shows that this kind of diversity could be mere taxonomic noise: what matters are the causal relationships—the ecological relationships—that sustain a stable holobiont. This is then seen as a succession of steps that can be fulfilled by different *things*, instead of a collection of specific *things*. What was needed, therefore, was a process-based explanation for holobionts.

The Turnbaugh et al. paper stimulated many evolutionary biologists to think about ecosystems (of which holobionts are an example) more broadly. The notion that microbiomes are more conservative functionally than taxonomically is common now, although "function" and taxonomy cannot be measured in the same units, really, so it is deeply unsensible to say that one is more conservative than the other (Inkpen et al., 2017). But measurement of function and taxonomy aside, the idea that what is conserved by natural selection is the pattern of functional interactions—the process or "song"—is the essence of ITSNTS. This seemed a welcome abstraction from the taxonomic (lineage) constancy that the hologenome theory of evolution must assume if it asserts that hologenomes, i.e., collections of genomes (i.e. specific *things*), are classical "units of selection." That assumption is often belied empirically.

Hence the need for an alternative such as ITSNTS. And already in 2010, Doolittle and Zhaxybayeva (2010. p. 110) wrote that the work of Turnbaugh et al.:

> … suggests a radical model for understanding the [philosophical] status of communities, in which they are seen as "real" – as discrete spatiotemporal entities, targets of selection and categories within the hierarchy of biological organization – perhaps more real than the phylogenetic lineages or "species" that are their members. In other words, it may be community composition in terms of functional genes that is preserved by selection, not the phylogenetic identity of the bearers of the genes; the song, not the singer, matters.

In ITSNTS theory, the "unit of selection" was thus cast as the process (the "song") implemented by the taxa (the "singers") which performed its steps. Songs don't reproduce, but are re-produced, whenever taxa which can perform their steps are brought together. Many taxa have that ecological ability because of close phylogenetic relationship or lateral gene transfer. This can be true even if those taxa are assembled at different times and in different places, as in many biogeochemical cycles. The steps in a song, if incompletely cyclized, allow intermediate compounds to accumulate and this encourages the evolution of taxa that can use them. This is called "recruitment" in ITSNTS theory. "Because there's a song [with accumulating compounds], there [can be new] singers: because there are [new] singers, there's a [dynamic] song" is one way to summarize the reciprocal causation involved in the dynamics of *recruitment*. In any case, a cyclical form is best, because it allows all intermediate components to be regenerated (Doolittle & Inkpen, 2018).

But a common criticism of ITSNTS is that it involves an uncomfortable mix of things and processes. What are the processes that behave as units of selection, and how do they interact with the things that make them up? These criticisms are strengthened by the belief that processes are immaterial while things are material. Such beliefs are often voiced by scientists, who see the entities making up the consortia as somehow more "real" than the patterns of interaction between them. It is also voiced by some philosophers, who embrace a dualist perspective (e.g. Robinson and Weir, 2026) but worry about how there can be causal connections between domains.

We think that these criticisms of ITSNTS (and process philosophy more broadly) are unfortunate because unfounded. They falsely superimpose the process-thing distinction over the immaterial-material distinction. The processeses at the heart of ITSNTS are material entities in the same way that reproduction is a material process. Processes need not be immaterial. The process ontology of Dupre, by arguing that all entities are in fact processes, albeit material, can help make this point, as we do in the next section. Epistemically, it could *still* be that some phenomena are best *explained* as the interactions between the properties of things, and some phenomena are better *explained* as processes simpliciter. We think ITSNTS emphasizes the latter over the former, *and that is its novelty, as an approach to holobiosis*. ITSNTS simply calls for a better intergration of evolutionary biology and ecology explanations.

6.5 ITSNTS: Evolutionary Biology and Ecology United

As we've said, we don't propose here to decide whether the world is fundamentally made up of *things*, their dispositional properties determining what happens (in terms of their interactions with other things), or instead fundamentally made up of *processes*, things just being how these processes manifest themselves at a particular time. We suspect that anything can be translated from one language into the other: it may just be a question of how *difficult* that translation is, and of course this difficulty is a product of language, of the state of our knowledge construction enterprises and their evolution (Foucault, 1966; Quine, 1951, 1968).

Dupré and Nicholson (2018) argue rightly that the history of both Western philosophy and science has been dominated by thing- or substance-ontology. They suggest that at least for (philosophy of) biology, this is problematic. For instance, they claim that:

> You can leave your typewriter in an empty loft and return a decade latter and start using it again. But if you accidentally leave your hamster in the loft, you will not have a hamster for very long. (Dupré & Nicholson, 2018, p. 15)

The obviousness of this contrast may be misleading. Some cryophilic bacteria have metabolisms as slow as the decay of any typewriter ribbon, and yet they are "living", while the notion of a hamster as an individual living thing may be outdated (Gilbert et al., 2012). Some of its microbiome outlast even the typewriter itself. Thinking that living things have a quality that non-living things don't is close to vitalism, after all. Still, a process account seems more unavoidable for living things, like Dupré and Nicholson's hamster, than for non-living entities like their typewriter, and ITSNTS was intended to be a theory uniting the study of change in the biological and non-biological realms, for all that there may really be no difference in the end (Mariscal & Doolittle, 2020). Perhaps a thing, to not be a process in an epistemically-driven perspective, needs for most of its pieces to be eternal (for as long as is relevant to the scale of the inquiry). In contrast, a process is characterized by its function (causal effect) and the function (causal effect) of its parts, irrespective of the stability of its composition. As already mentioned, most systems can probably be viewed in both lights. What we wish to emphasize is that holobiosis gains to be studied as a process, wherever it lies on the ontological continuum that relates things like typewritters and processes like hamsters.

Certainly we regard ourselves epistemically as "the same" individuals as we were as babies, and species, by currently predominate accounts, are individuals, too. And yet few of the atoms which make us up, and arguably only a tiny and random fraction of the atoms which make up *Homo sapiens* now are the very same atoms that we, or *H. sapiens*, started with. We, and the species to which we belong, are dynamic, changing entities, better explained as ecological processes than as static things. As Dupré and Nicholson argue, the metabolisms, physiology, life cycle, genetics, ecology and evolution of both are conserved as processes more than are the things that implement them. Species and organisms can still be conceived of as individuals, but they are individual *processes*, i.e., songs.

And indeed, Bapteste and Papale (2021) have suggested that songs, in the sense of ITSNTS, can be found at every level of the biological hierarchy. Any entity that we used to describe as a thing can be modeled in terms of the interactions between its components. If the focus is on these patterns of interactions and their dynamics, then the thing "becomes" a process. This means that the process ontology inherent in ITSNTS explanations can be generalized in a way that is relevant to evolutionary biology, epistemically. Entities such as ecosystems, multispecies consortia, the Earth system, biogeochemical cycles, sets of autocatalytic chemical reactions, as well as more classical "things" such as organisms, species or genes, can thereby be explained as processes and candidate units of selection, more broadly construed, as in ITSNTS

(Doolittle & Inkpen, 2018). This takes us far away from the classical view of the unit of selection, yet it provides an evolutionary sound perspective of these processes.

The thing/process dichotomy translates a difference of perspective that is often both theoretically and empirically relevant. Whether a researcher approaches an organism as a singer or a song, his or her work will be influenced by the underlying metaphysical choice, as is always the case. For instance, researchers interested in multispecies consortia as things might focus on their genetic and taxonomic constitution, while treating them as songs might lead the same researchers to focus on patterns of interactions.

But in both cases, we are talking about material entities: processes are the dynamic flow with which we interact, as researchers. Abstracting away their dynamics to focus on entities as static things does not increase their materiality, it simply turns the coin around. The coin, whether you look at one side or the other, is a material entity. By adopting process ontology here articulated as an epistemic matter first and foremost, ITSNTS avoids the critique of some of its detractors according to which it requires immaterial entities to interact with material ones. It may treat these entities as epistemically different, but it assumes that ontologically speaking, they are the same, i.e., material entities.

6.6 The Classical View of ENS Versus an Extended Replicator/Interactor View

A question which remains from the above analysis, and about which we cannot help comment, is whether some general theories of evolution by natural selection are more apt for things, and others more apt for processes. As discussed above, the issue with routine hologenome thinking is not just that it focuses on the wrong unit, but the wrong kind of unit: a thing rather than process. But it also employs a classical Lewontin's Recipe approach to ENS—it attempted to treat the hologenome as a unit of selection in this sense. In contrast, ITSNTS is really a species of a different approach, more aligned with philosopher David Hull's interactor-replicator theory, and also—when interactors have too many parents—more aligned with ecology. To end, we'll try to explain why this might matter.

Godfrey-Smith, in his 2009 *Darwinian Populations and Natural Selection*, presents a one-part theory of ENS, and argues (as well-articulated there and in Godfrey-Smith, 2015) that: "benefits to life as a whole are by-products—they're accidental. The interactions between species are consequences of the traits and behaviours that evolutionary processes within those species give rise to, and those processes are driven by reproductive competition within each species." This perfectly illustrates a thing-based approach to ENS.

We'd argue, as would Godfrey-Smith, that this classical view is readily extended to encompass multilevel selection theory, in which any thing-like entity "genes, cells, social groups and species can all, in principle, enter into change of this kind"

(Godfrey-Smith 2012, p. 2161) if and only if such entities exhibit something like differential reproduction with heredity. Species, for instance, do this by speciating.

On the other hand, David Hull, in his much earlier (1980) essay presenting his replicator/interactor theory holds that,

> When biologists address the issue of the levels at which selection can occur, they take the traditional organizational hierarchy as fundamental and the level at which selection operates as a variable. As a result, selection wanders from one level to the next from time to time and from group to group, sometimes acting on genes, sometimes organisms, sometimes colonies, etc. As long as the traditional hierarchy is taken as basic and the levels at which selection can occur as variable, no simple, nomothetic generalizations are likely to materialize. Rather, the evolutionary *process* must be taken as basic and the levels defined in terms of it. Two entities that perform the same function in the evolutionary process must be classed as the same sort of entity even if one happens to be an organism and the other a colony or a population. (emphasis ours, Hull 1980, p. 316)

We'd submit that Godfrey-Smith's classical formulation continues with the emphasis of most biologists on things, as illustrated by the citation above, specifically things that reproduce. It is members of Darwinian populations that are the "units of selection", and the process of ENS involves focusing on the units as things according to this view. Its currency is the reproductive lineages of reproducers: thing-like parents and thing-like offspring. Hull's replicator/interactor formulation, on the other hand, thinks of evolution as a *process* (see our emphasis) and defocuses things, at least insofar as interactors are not necessarily that: they are songs. According to Hull, ENS occurs when the differential extinction and proliferation of *interactors* (entities that interact with their environments) cause the differential perpetuation of *replicators* (entities that are copied reliably from generation to generation).

And in our extended form of replicator/interactor thinking—for instance about biogeochemical cycles, like that for nitrogen—the emphasis on processes and patterns of interaction is even more important than the emphasis on things (Inkpen and Brunet, under review; Papale and Doolittle, 2024). Contrary to how Hull first theorized them, we believe interactors don't need to be *cohesive*. That's what ITSNTS was about: the recognition that the thing-thinking underlying the hologenome theory of evolution could (and maybe should) be replaced by the process framework of ITSNTS. Just a shift in emphasis (a gestalt shift?, maybe), leading to a disambiguation and redefinition of units of selection, as exemplified in Suárez and Lloyd (2023), and before that in Lloyd 's work (see Lloyd (2024) for a summary, have argued. This was the motivation behind the formulation of ITSNTS.

The paleoecologist Peter Roopnarine (2025) has formulated an empirically based theory, similar to ITSNTS. He argues:

> … that: (1) EFNs [ecological functional networks] belong to a class of entities above the species level, that includes clades, ecosystems and function interaction networks … (2) that geological persistence arises from selection acting on EFN properties which vary according to variance within and among communities; and (3) that EFNs represent evolved functional networks.

And, notably, he also, like we, interprets his theory as being more in synch with Hull's replicator/interactor framework than Godfrey-Smith's classical version of ENS. He says:

According to Hull's replicator-interactor framework of evolution by natural selection (Hull, 1980), interaction with the environment may take place at a level of organization above that of replicators, with replicators causing the interaction which in turn affects replication. EFNs are interactors, and species are both interactors and replicators. It is unnecessary, however, for EFNs to be replicating units with heritable traits to evolve, for selection here is not causing differential reproductive success, but instead differential persistence ....

So, replicator/interactor framework makes all of Life—all of its so-called hierarchical structure—subject to Darwinian principles, not just the organisms and groups of organisms on which Richard Dawkins focused in *The Selfish Gene*. (He broadened that perspective in *The Extended Phenotype*, extending the selfishness of genes as replicators.) These principles are, in our view, just those of "universal Darwinism" (Dawkins, 1983), and our concept of replicator has loosened and expanded as well. We see as "replicators" any *reproducing* things which are subject to Lewontin's Recipe (provided that they don't have "too many parents") and fit into its expansion into multilevel selection theory (Doolittle, 2024). Genes, organisms, populations of organisms and species do fit in, but horizontally transmitted holobionts, ecosystems and clades don't. And we see as relevant interactions of the interactor, not only responses to the physical (abiotic) environment but also what are often called direct and indirect interactions between species or individuals making the interactors up. Still, interactors evolve, increasing in frequency and for reasons, and we are tempted to call that natural selection as well (Papale and Doolittle, 2024). So, replicator/interactor-thinking does seem more apt for explaining the evolution of processes than does Lewontin's recipe.[1]

6.7 Conclusion

We think that Dupré and Nicholson have correctly pointed out that science (or at least evolutionary biology) has been very thing-oriented, to the neglect of processes thinking. This characterization is true at both epistemological and ontological levels. But Lewontin's Recipe seems more about things and their properties as explanation than does the replicator/interactor framework of Richard Dawkins and David Hull, which opens the door more readily to processes and ecology. So there might indeed be a spectrum, epistemically, which ITSNTS straddles uncomfortably. But it's not clear to us why the world out there cannot be populated by entities that are really

[1] Note that at least one of us believes that the classical approach to ENS can be adapted to processes. Yet all of us agree that Godfrey-Smith and other contemporary proponents of that approach have tended hold on to the perspective according to which units of selection are to be conceptualized as static things rather than dynamic processes. That fails to capture the inherent complexity of evolutionary processes, as emphasized in Papale et al. (2024) and by the ITSNTS theory.

neither, and that both descriptions (things or processes) are the best that our evolved minds can do (Schwitzgebel, 2024).

References

Bapteste, E., & Papale, F. (2021). Modeling the evolution of interconnected processes: It is the song and the singers. *BioEssays, 43*, 2000077.

Brunet, T. D. P. (2024). Local ontology: Reconciling processuralism and new mechanism. *European Journal for the Philosophy of Science, 14*, article 28.

Dawkins, R. (1983). Universal darwinism. In D. S. Bendell (Ed.), *Evolution from molecules to man.* Cambridge University Press.

Doolittle, W. F., & Zhaxybayeva, O. (2010). Metagenomics and the units of biological organization. *Bioscience, 60*, 102–112.

Doolittle, W. F., & Booth, A. (2017). It's the song, not the singer: An exploration of holobiosis and evolutionary theory. *Biology and Philosophy, 32*, 5–24.

Doolittle, W. F., & Inkpen, S. A. (2018). Processes and patterns of interaction as units of selection: and introduction to ITSNTS thinking. *Proceedings of the National Academy of Sciences USA, 115*, 4006–4014.

Doolittle, W. F. (2024). *Darwinizing Gaia: Natural selection and multispecies community evolution.* MIT Press.

Douglas, A. E., & Werren, J. H. (2016). Holes in the hologenome: Why host-microbe symbioses are not holobionts. *mBio, 7*. 10.1128/mbio.02099-15.

Dupré, J. (2017). The metaphysics of evolution. *Interface Focus, 7*, 20160148.

Dupré, J., & Nicholson, D. J. (2018). A manifesto for a processual philosophy of biology. In D. J. Nicholson & J. Dupré (Eds.), *Everything flows: Towards a processual philosophy of biology.* Oxford University Press.

Foucault, M. (1966). *Les mots et les choses.* Éditions Gallimard.

Gilbert, Scott F., Sapp, j., & Alfred I. Tauber. (2012). A Symbiotic View of Life: We Have Never Been Individuals. *The Quarterly Review of Biology 87*(4), 325–41. https://doi.org/10.1086/668166.

Godfrey-Smith, Peter. (2009). *Darwinian Populations and Natural Selection.* Oxford University Press.

Götzenberger, L., de Bello, F., Bråthen, K. A., Davison, J., Dubuis, A., Guisan, A., ... & Zobel, M. (2012). Ecological assembly rules in plant communities—approaches, patterns and prospects. *Biological reviews, 87*(1), 111-127. https://doi.org/10.1111/j.1469-185X.2011.00187.x.

Hull, David L. (1980). "Individuality and Selection." *Annual Review of Ecology and Systematics 11*, 311–32.

Inkpen, S. A., Douglas, G. M., Brunet, T. D. P., Leuschen, K., Doolittle, W. F., & Langille, M. G. (2017). The coupling of taxonomy and function in microbiomes. *Biology & Philosophy, 32*(6), 1225–1243. https://doi.org/10.1007/s10539-017-9602-2.

Javier, Suárez Elisabeth A., Lloyd (2023). *Units of Selection Cambridge University Press.*

Lewontin, R. C. (1970). The units of selection. *Annual review of ecology and systematics, 1*-18.

Linquist, S. (2019). Why ecology and evolution occupy distinct epistemic niches. *philosophical topics, 47*(1), 143-166.

Lloyd, Elisabeth. (2024). "Units and Levels of Selection." In The Stanford Encyclopedia of Philosophy, Summer 2024, edited by Edward N. Zalta and Uri Nodelman. Metaphysics Research Lab, Stanford University. https://plato.stanford.edu/archives/sum2024/entries/selection-units/.

Madhusoodanan, J. (2019). Do hosts and their microbes evolve as a unit?. *Proceedings of the National Academy of Sciences, 116*(29), 14391–14394. https://doi.org/10.1073/pnas.1908139116.

Mariscal, Carlos, and W. Ford Doolittle. (2020). "Life and Life Only: A Radical Alternative to Life Definitionism." *Synthese*197(7): 2975–89.
Moran, N. A., & Sloan, D. B. (2015). The hologenome concept: helpful or hollow? PLoS biology, 13(12), e1002311. https://doi.org/10.1371/journal.pbio.1002311.
Morris, J. J. (2018). What is the hologenome concept of evolution?. *F1000Research*, 7, F1000-Faculty.
Morris, M. M., Kimbrel, J. A., Geng, H., Tran-Gyamfi, M. B., Yu, E. T., Sale, K. L., & Mayali, X. (2022). Bacterial community assembly, succession, and metabolic function during outdoor cultivation of Microchloropsis salina. MSphere, 7(4), e00231-22. https://doi.org/10.1128/msphere.00231-22.
Papale, F., &, W. F. (2024). "Towards a More General Theory of Evolution by Natural Selection: A Manifesto." *Philosophy, Theory, and Practice in Biology 16*(1).
Quine, Willard V. 1951. Ontology and Ideology. *Philosophical Studies: An International Journal for Philosophy in the Analytic Tradition 2*(1): 11–15.
Quine, W. V. O. 1968. Ontological Relativity. *Journal of Philosophy 65*(7): 185–212. https://doi.org/10.2307/2024305.
Queller, David C., and Joan E. Strassmann. 2016. Problems of Multi-Species Organisms: Endosymbionts to Holobionts. *Biology & Philosophy 31*(6): 855–73. https://doi.org/10.1007/s10539-016-9547-x.
Roopnarine, P. D. (2025). Selection, evolution and persistence of paleoecological systems. *Frontiers in Earth Science*, 13, 1528448.
Robinson, Howard, and Ralph Weir. 2026. Dualism. In The Stanford Encyclopedia of Philosophy, Spring 2026, edited by Edward N. Zalta and Uri Nodelman. *Metaphysics Research Lab, Stanford University*. https://plato.stanford.edu/archives/spr2026/entries/dualism/
Schwitzgebel, E. (2024). *The Weirdness of the World*. Princeton: Princeton University Press.
Simons, P. (2018). Processes and precipitates. Everything flows. *Towards a processual philosophy of biology*, 49–60.
Skillings, Derek. 2016. Holobionts and the Ecology of Organisms: Multi-Species Communities or Integrated Individuals? *Biology & Philosophy 31*(6): 875–92. https://doi.org/10.1007/s10539-016-9544-0.
Tian, Liang, Xu-Wen Wang, Ang-Kun Wu, et al. 2020. Deciphering Functional Redundancy in the Human Microbiome. *Nature Communications 11*(1): 6217. https://doi.org/10.1038/s41467-020-19940-1.
Turnbaugh, Peter J., Micah Hamady, Tanya Yatsunenko, et al. 2009. A Core Gut Microbiome in Obese and Lean Twins. *Nature 457*(7228): 480–84. https://doi.org/10.1038/nature07540.
Tyler, Volk Stephen H., Schneider James R., Miller Eileen, Crist Penelope J., (2014) Boston Scientists Debate Gaia The Next Century Gaia Is Life in a Wasteworld of By-products The MIT Press 27-36
Zilber-Rosenberg, I., & Rosenberg, E. (2008). Role of Microorganisms in the Evolution of Animals and Plants: The Hologenome Theory of Evolution. *FEMS Microbiology Reviews 32*(5), 723–35. https://doi.org/10.1111/j.1574-6976.2008.00123.x.

Chapter 7
Promiscuous Mutualisms: John Dupré and the Centrality of Symbiosis in the Living World

Thomas Pradeu

Abstract My goal in this chapter is to show that symbiosis, especially symbiotic interactions between 'macrobes' (non-microbial organisms) and microbes, is a central topic in John Dupré's work. This topic is undeniably one of the topics by which Dupré has gained wide recognition among practicing scientists, in addition to his recognition within the philosophical community. I spell out why symbiosis has turned out to be a sort of ideal trope for Dupré, as it encapsulates many of his main personal themes—from promiscuous realism to process ontology. I then show that immunology offers one interesting and complementary entry point into microbial symbiosis and biological individuality.

7.1 Symbiosis in Dupré's Work

7.1.1 Dupré, from a Bibliometric Viewpoint: Microbes and Microbial Symbioses as a Central Topic, Recognized by Practicing Scientists

In this section, I present evidence that John Dupré, in addition to his compelling recognition as one of the most prominent philosophers of science in the last five decades or so, has also gained wide recognition among scientists. Moreover, I explain why his work on microbes, especially on host-microbe symbiosis, offers a major example of this recognition by scientists.

Let's start with a broad question: To what extent has John Dupré's work been visible to practicing scientists? Of course, being visible to scientists is not the only goal a philosopher of science may have. Some philosophers of science consider that being visible to scientists is very secondary, or even not desirable at all. Yet for

T. Pradeu (✉)
ImmunoConcept (UMR5164), CNRS & University of Bordeaux, France, Presidential Fellow, Chapman University, Orange, USA
e-mail: thomas.pradeu@u-bordeaux.fr

S. Leonelli et al. (eds.), *Life Beyond Human Boundaries: John Dupré's Philosophy and the Interdisciplinary Study of the Life Sciences*,
https://doi.org/10.1007/978-3-032-26977-5_7

philosophers of science who, like Dupré, practice a scientifically engaged form of philosophy of science, such recognition matters, because it signals that their work as philosophers has been noticed and discussed by scientists. This is one key dimension of what I have proposed to call 'philosophy *in* science' (Pradeu et al., 2024). Although visibility can be achieved in many different ways, its clearest and most convincing form consists in being cited in scientific journals (Khelfaoui et al., 2021). Indeed, when a philosopher of science is cited in scientific journals, it means that their work has been deemed worthy of attention by certain scientists, who use that work in their own research. Such work written by philosophers of science and cited by scientists can be published either in philosophy journals or in science journals, although the second option significantly increases the probability of being cited by scientists (Pradeu et al., 2024).

From the bibliometric viewpoint of citation analysis, Dupré has, overall, been cited 538 times according to the Web of Science (accessed in April 2025). The best Dupré's 'vintages' were 2010 in terms of number of publications (Dupré published 7 papers that year) and 2011 in terms of number of citations per year (Dupré received 48 citations that year). On average, papers by Dupré gather 17.5 citations. Of note, among Dupré's many papers, only 34 appear in Web of Science. This is because Web of Science has a notoriously poor coverage of the humanities, while it is considered a reliable and demanding ressource for science and technology. The rich but nonetheless not too inclusive coverage of Web of Science (in particular, it excludes most so-called predatory or quasi-predatory journals, although the database of Web of Science and, even more so, that of Scopus, have been shown to be imperfect from this point of view (Frandsen, 2017)) makes it a particularly appropriate tool for the question we are asking here, namely Dupré's visibility to scientists.

So, if we now focus on Dupré's visibility *within scientific journals*, Dupré has been cited 239 times in science journals in total. Please keep in mind that, for a philosopher, achieving visibility in science is extremely difficult, mainly because of the traditional disciplinary 'silos' of research. Among the scientific papers that cite Dupré, the scientific domains in which Dupré has been highly cited are, in this order, 'Microbiology' (about 21% of total number of citations), 'Genetics & Heredity' (about 20%), 'Evolutionary Biology' (about 20%), and 'Biochemistry Molecular Biology' (about 16%). Unsurprisingly, the scientists who have cited Dupré the most are Eric Bapteste, Eugene Koonin, and Purificacion Lopez-Garcia. Dupré's most cited papers in science journals according to Web of Science appear in Table 7.1.

A clear lesson here is that, among Dupré's papers cited by scientists, those related to microbiology and the impact of microbes on 'macrobes' (i.e., 'big', non-microbial, organisms) stand out. This is the case of five of the six most cited papers (Bapteste et al., 2009; Bapteste & Dupré, 2013; Dupré, 2010; O'Malley et al., 2010; O'Malley & Dupré, 2007a). Among these 5 papers, 3 have been co-written with Maureen O'Malley, one of the most prominent philosophers of microbiology internationally. Two of these papers have been co-written with scientists, especially with Eric Bapteste, who holds two PhDs, one in biology and one in philosophy. Finally, one paper was published in a scientific journal, which is quite a frequent thing among scientifically-inclined philosophers and generally increases the probability of being

Table 7.1 Dupré's most cited papers in science journals according to Web of Science

Title	Authors	*Journal*	Year	Number of citations in science journals
Prokaryotic evolution and the tree of life are two different things	Bapteste, Eric; O'Malley, Maureen A.; Beiko, Robert G.; Ereshefsky, Marc; Gogarten, J. Peter; Franklin-hall, Laura; Lapointe, Francois-Joseph; dupre, John; Dagan, Tal; Boucher, Yan; Martin, William	*Biology direct*	2009	140
Size doesn't matter: Towards a more inclusive philosophy of biology	O'Malley, Maureen A.; dupre, John	*Biology & Philosophy*	2007	24
Towards a processual microbial ontology	Bapteste, Eric; dupre, John	*Biology & Philosophy*	2013	14
Disciplinary baptisms:: A comparison of the naming stories of genetics, molecular biology, genomics, and systems biology	Powell, Alexander; O'Malley, Maureen A.; Mueller-Wille, Staffan; Calvert, Jane; dupre, john	*History and philosophy of the life sciences*	2007	11
The tree of life: Introduction to an evolutionary debate	O'Malley, Maureen A.; Martin, William; dupre, John	*Biology & Philosophy*	2010	10
The polygenomic organism	Dupre, John	*Sociological review*	2010	8

cited by scientists (Pradeu et al., 2024)—an effect which is quite striking in Table 7.1. Three papers have been published in *Biology & Philosophy*, which, for a philosophy of science journal, is on average relatively visible to scientists (Khelfaoui et al., 2021). Crucially, if one examines the scientific papers that cite Dupré's papers, we see that several of them are themselves both highly cited and written by scientists who are central in their own field, which confirms the recognition of Dupré's work in these domains (e.g., (Mueller et al., 2012), in which Bill Martin is the senior author, has been cited 560 times according to Web of Science).

Thus, we can conclude that Dupré has gained significant visibility in science and that the topic of microbes, especially macrobe-microbe symbiosis, has been one of the most important sources of this visibility in science. Let's now explore more specifically what Dupré has said about symbiosis and on this relates to what other philosophers have written on this topic.

7.1.2 Philosophers on the Topic of Symbiosis

Dupré has been one of the leading philosophers working on the fascinating issue of symbiosis. It is no wonder that the topic of symbiosis has attracted many philosophers, as it raises major conceptual, theoretical, and philosophical questions (Suárez, 2018). These questions include biological individuality (Booth, 2014; Clarke, 2011; Parke, 2021; Parke et al., 2018; Pradeu, 2011), including through the revisiting of the concept of an 'interactor' (Lloyd & Wade, 2019) and potentially challenging the concept of 'Darwinian individuals' (Godfrey-Smith, 2013). It also gave rise to legion of discussions on whether host-microbiome associations should be considered 'holobionts' and units of selection (Baedke et al., 2020; Chiu & Gilbert, 2015; Queller & Strassmann, 2016; Roughgarden et al., 2017; Skillings, 2016; Suárez, 2020; Suárez & Stencel, 2020; Triviño & Suárez, 2020). Symbiosis also served as a basis to contribute to question traditional approaches to fitness and to promote the idea that fitness may be more about persistence than about reproduction (Bouchard, 2009, 2010). A related question is whether symbiosis challenges our traditional concept of lineages (Dupré & O'Malley, 2009; Haber, 2016). Other widely discussed questions include whether symbiosis forces us to revisit certain traditional approaches to development, including gene-centric and internalist perspectives (Gilbert, 2002, 2014; McFall-Ngai, 2002; Pradeu, 2011), and whether health depends on a certain composition or a certain balance or 'homeostasis' of the microbiome (Hooks & O'Malley, 2017; Kelley, 2024; O'Malley, 2024). Dupré has participated in all of these debates. Interestingly, the work on symbiosis done by philosophers, including Dupré, has attracted the attention of leading scientific experts working on symbiosis, including Angela Douglas, Margaret McFall-Ngai, Rob Knight, Gérard Eberl, Scott Gilbert, Ford Doolittle, Seth Bordenstein, Eric Bapteste, among many others. These scientists have cited and used work done by philosophers, and several of them have co-authored papers with philosophers.

7.1.3 Dupré on the Topic of Symbiosis, and why it Is Central in his Work

Dupré stands out as one of the most central authors in the HPS literature on symbiosis and the microbiome, most often through his long-lasting and extremely fruitful collaboration with Maureen O'Malley. Indeed, this partnership has generated many papers on a considerable diversity of topics—ranging from system biology to characterization of living things and to metagenomics, the metaphor of the tree of life, philosophy of microbiology, and the naming of emerging disciplines (Bapteste et al., 2009; Dupré & O'Malley, 2007, 2009; O'Malley et al., 2010; O'Malley & Dupré, 2005, 2007a, 2007b; Powell et al., 2007). In my view, this collaborative work, especially in so far as it directly addresses symbiosis, results from the felicitous and extremely productive combination of expertise of two outstanding philosophers of biology—one, O'Malley, with cutting-edge knowledge of the most central debates in microbiology, and the other, Dupré, with a number of philosophical ideas or 'obsessions' (I mean this in a very positive way) that found in microbiology their best possible concrete realization.

In other words, the topic of symbiosis has been for Dupré's work a fantastic nexus, putting together recent scientific discoveries and critical philosophical questions. In fact, many of these scientific discoveries have been a blessing for Dupré, in so far as they came to confirm and illustrate some of his most central preexisting philosophical ideas. This is particularly clear with two major topics of Dupré's philosophical work pertaining to metaphysics and epistemology, namely the *plurality of legitimate classifications* and the *interpenetration of living processes*. Indeed, the demonstration that symbiosis is ubiquitous in the living world can first be interpreted as a very concrete realization of Dupré's "promiscuous realism" as far as the topic of classification is concerned. Promiscuous realism, spelled out in Dupré's magnum opus *The Disorder of Things* (Dupré, 1993), proposes that there exist in most biological contexts different ways to classify objects that are equally legitimate, while no single classification system captures all the relevant relationships or serves all scientific purposes. This has led to the key idea that realism and pluralism are not necessarily in tension. The omnipresence of symbiotic relationships in nature strengthens promiscuous realism in several ways. First, the demonstration that symbiosis is ubiquitous in the living world blurs traditional boundaries and shows that the interpenetration of biological objects is not just conceivable but is in fact the norm in nature. This challenges our ability to neatly categorize organisms as distinct individuals. Consider lichens, for instance. A lichen is a composite organism that emerges from the symbiosis between a fungus and algae or cyanobacteria. Such associations resist straightforward classification within traditional taxonomic frameworks, and overall it is not easy to say to which 'natural kind' a lichen belongs (Spribille et al., 2022). Second, work on symbiosis shows that different perspectives and delineations can co-exist and be legitimate: for instance, a human body can be legitimately viewed as a single organism or as a complex ecosystem of human and microbial cells, depending on the scientific and/or medical question being asked. This is related to the third point demonstrated by symbiosis, which is context-dependency: How we classify symbioses based on traditional ecological relationships (typically mutualism, commensalism, parasitism) often depends on the system considered, timescale, and which aspects of the relationship one decides to emphasize (Méthot & Alizon, 2014). Fourth, work on symbiosis strengthens the idea that essentialism must be rejected: symbiotic interactions show how biological entities lack fixed 'essential natures' and are instead dynamically constituted through their relationships and interactions with other entities. In a nutshell, recent work on the ubiquity of symbiosis provides precise and compelling evidence in favor of Dupré's promiscuous realism by showing how biological reality often resists our attempts to impose neat categorical boundaries.

Studies on symbiosis show that organisms and lineages fuse and separate on a very regular basis. In many cases, organisms are local and temporary concentrations of multiple intertwined and interacting components that belong to multiple categories. Of course, the degree of interpenetration of entities such as organisms has led

Dupré to question the very notion of 'entity' and its relevance in today's biological sciences. Dupré has proposed that an ontology and epistemology of processes should replace the traditional ontology and epistemology of things/entities (Dupré, 2012; Nicholson & Dupré, 2018). My hypothesis, which would require much more detailed work and explanations, is that symbiosis, together with later gene transfer, has been for Dupré a major argument and motivation for introducing process ontology (see (O'Malley & Dupré, 2007a)).

Overall, then, the topic of symbiosis, especially between 'macrobes' and their microbiomes, has had a crucial impact on Dupré's work in the last two decades from at least three points of view:

1. Symbiosis challenges traditional organism boundaries: Dupré argues that the discovery of the ubiquity and importance of microbiomes fundamentally challenges our conception of organisms as discrete, autonomous individuals. This idea is well illustrated by the concept of 'polygenomic organism' proposed by Dupré. According to this concept, organisms must be reconceptualized as fundamentally polygenomic and symbiotic communities rather than genetically homogeneous individuals (Dupré, 2010).
2. Symbiosis offers one of the best arguments in favor of a process ontology, participating in showing the dynamic, interconnected nature of living systems (Nicholson & Dupré, 2018).
3. Symbiosis offers a ubiquitous and radical illustration of promiscuous realism, by showing the concrete coexistence of several distinct but equally relevant ways to carve up biological realities.

7.2 A Quick Account of why the Immune System Offers One Interesting and Complementary Entry Point into Symbiosis

As certainly all the participants to Dupré's *Festschrift* at Egenis in 2024 and all the contributors to this volume, my work has been influenced by Dupré's work. It might be relevant here to emphasize one convergence and one inspiration. The convergence concerns biological individuality. The inspiration concerns process thinking.

As far as the topic of biological individuality is concerned, Dupré and I have come from very different perspectives (Dupré's approach was based primarily on evolution, classification, and later microbiology; mine was based mainly on immunology and physiology), but we came to very similar—and, it seems to me, complementary—conclusions. First, we have both independently argued that symbiotic relationships and microbiomes fundamentally challenge the traditional view of organisms as autonomous, genetically homogeneous individuals (Dupré, 2010; Pradeu, 2010). Second, we have both proposed to think about biological individuality from the point of view of biological approaches focusing on the 'how?' as well as biological approaches focusing on the 'why?' (i.e., focusing on what Mayr had dubbed, somewhat controversially and inappropriately, 'functional biology'(Mayr, 1961)). Previous philosophical investigations on biological individuality had focused on evolutionary considerations, neglecting the fact that physiological and metabolic considerations could usefully complement these evolutionary considerations (Bouchard & Huneman, 2013; Buss, 1987; Hull, 1978, 1980, 1992; Maynard Smith & Szathmáry, 1995; Michod, 1999). (This is not surprising given the general tropism of philosophers of biology for evolutionary biology (Pradeu, 2017)). In their major paper "Varieties of living things" (Dupré & O'Malley, 2009), Dupré and O'Malley insist that we should reconceptualize living things as collaborative associations of different types of entities with varying degrees of independence. Life should be understood as operating at the intersection of metabolic processes and lineage-forming entities. Collaborative organisms should be understood as multilineage systems that function together as metabolically integrated wholes. In parallel, in my PhD dissertation written under the supervision of Jean Gayon at Panthéon-Sorbonne University and defended in 2007, I proposed that physiological domains, especially immunology but other domains as well, could offer a useful, and, contrary to what Hull had suggested (Hull, 1992), theoretically very rich, perspective on biological individuality (see also (Pradeu, 2010, 2016)). This convergence between our views has been picked up by other philosophers who have written on biological individuality, most prominently Peter Godfrey-Smith, who built on this combination between physiology and evolutionary biology in his work on this topic (Godfrey-Smith, 2013, 2014).

The inspiration concerns process thinking. Although I have been agnostic on the utility of an *ontology* of processes, I think that an *epistemology* of processes can definitely be relevant and useful. Inspired by Dupré's work, I have tried to show that the ubiquity of symbiosis in the living world offered a major argument in favor of '*genidentity*'—a concept proposed by Kurt Lewin and expanded in different domains, including physics, by Hans Reichenbach (Lewin, 1922; Padovani, 2013; Reichenbach, 1956). Interestingly, David Hull, not just one of the most important philosophers of biology but also a central contributor to debates about biological individuality, saw the fruitfulness of the concept of genidentity. In particular, Hull wrote: "The idea that comes closest to capturing individuality in organisms and possibly individuals as such is genidentity. As its name implies, this criterion allows for change just as long as it is sufficiently continuous. The overall organization of any entity can change but it cannot be disrupted too abruptly" (Hull, 1992). In fact, Hull's essential diagrams about biological individuality back in his fouding paper published in 1978, where both species and organisms can split or fuse in very similar ways (Hull, 1978), were already a remarkable, if implicit, illustration of the genidentity view. In my (modest) contribution to process thinking, I insisted on an epistemology of processes that would incorporate traditional and newer lessons coming from a genidentity-based view. Building on work done with my colleague and friend Alexandre Guay in philosophy of physics, I suggested that the ubiquity of symbiosis strengthens the genidentity view (Guay & Pradeu, 2016; Pradeu, 2018) and makes several of Hull's diagrams (Hull, 1978) switch from "weird and rare" to “weird and almost omnipresent” in nature (on the consequences of this, see also (Haber, 2025)).

To better illustrate these convergences and inspirations, I will now offer a quick overview of why I have claimed that immunology was central in understanding biological individuality but also why I considered that the dominant self-nonself theory had to be replaced. I’m sure the reader will see the many similarities and echoes between Dupré’s work and mine on such questions.

First, I have tried to put forward a revised approach to biological individuality based on immunology. I have shown the key role of the concept of individuality in the birth of immunology as a scientific field in the 1920s and 1930s (Pradeu, 2012). Burnet’s self-nonself theory resulted from the unification of two sets of phenomena, pertaining to infectious diseases on the one hand and to transplantation on the other hand, under a single and powerful explanatory framework, i.e., the claim that immune systems have been selected for their capacity to detect and eliminate nonself (everything that is foreign to the organism under consideration). According to the self-nonself theory, an organism immunologically rejects everything that is foreign to it (the ‘nonself’), while it does not trigger an immunological response against its own constituents (the ‘self’) (on this history, see also (Tauber, 1994)). Although this framework has been extremely useful and still dominates today’s immunology, my main claim has been that the traditional self-nonself framework is wrong—for many reasons, one of the most important of which is the ubiquity of symbiotic interactions in nature, which shows that the interpenetration of ‘self’ and ‘nonself’ is pervasive and that the idea that every nonself is rejected by the immune system does not hold (Pradeu, 2012). Yet I have also argued that, despite the inadequacy of the self-nonself theory, immunology still has a lot to say about how boundaries of living systems are established and maintained, in highly complex, dynamic, and flexible ways. Because my goal was, very much in the spirit of ‘philosophy in science’ (Pradeu et al., 2024), to undermine the self-nonself framework *within immunology*, I tried to provide both a novel account of immunological individuality and a novel theoretical framework for immunology.

Given what we have just said, it is simply not true that immune systems work as systems of detection and elimination of nonself. There are too many exceptions, especially those gathered under the concept of “immunological tolerance”—including feto-maternal tolerance and tolerance to the microbiome (Pradeu & Carosella, 2006). This does not imply, though, that one should get rid of the idea that immunology is central in our understanding of biological individuality. The immune system plays an essential role in the construction and maintenance of biological individuality through its key activities of filtering over entry, filtering over presence, and plasticity (Pradeu, 2019). Through these activities, the immune system is involved in all the major dimensions of biological individuality: countability, delineate, persistence, cohesion, and uniqueness.

From a present-day immunological perspective, every organism can be conceptualized as a multi-species community, made of billions of eucaryotic as well as microbial cells. Yet it is essential to note that such a community has a very strong degree of unification—much stronger than most ecological communities in nature. The immune system plays a key role in this unification of these multiple constituents, both endogenous ('self) and exogenous ('nonself'). The immune system, by its filtering activities, does constitute a principle of inclusion and exclusion although, crucially, this discrimination is not based on the origin of the element under consideration: an element coming from the 'inside' of the organism can perfectly be eliminated (this happens all the time) and an element coming from the 'outside' of the organism can perfectly be tolerated, i.e., not immunologically rejected despite its 'nonself' nature (this happens all the time, too). This is what led me to propose the claim that every organism is an *immunologically unified chimera*—a complex and multispecies association, the boundaries of which are constantly and dynamically redrawn by the action of immune interactions. Moreover, this idea of 'heterogeneous individuality' applies almost universally, as immune systems have been discovered across all the living world (in bacteria, archaea, plants, animals) and everywhere they seem to display the capacity to not eliminate certain foreign entities, establishing instead long-term interactions with them.

In addition to this philosophical work on the role of immunology in defining biological individuality, I have tried to develop work in theoretical immunology. Here the goal has been to offer an alternative theoretical framework to the dominant self-nonself framework. In a nutshell, the discontinuity theory (Pradeu et al., 2013; Pradeu & Vivier, 2016) proposes that an effector immune response is triggered not by 'nonself' but instead by the sudden modification of the molecular patterns with which immune receptors interact. Simply put, immune systems have been selected through evolution for their capacity to detect sudden changes, regardless of the origins (self or nonself) of the changing element. This framework has many experimental and therapeutic consequences, including as far as host-microbiome interactions are concerned (Pradeu, 2012; Swiatczak & Cohen, 2015).

My goal here has simply been to illustrate how convergent and, I believe, complementary Dupré's perspective and mine have been. Although we had different backgrounds and focused on very different biological domains, we came to very similar conclusions on the role of symbiosis in redefining biological individuality and identity. I don't know if this is evidence in favor of the 'robustness' of this conclusion, but at least for today suffices to say that I hope that this is the case.

7.3 Conclusion

I have tried to show here that the topic of symbiosis, especially microbial symbiosis, has played an essential role in Dupré's thinking in the last three decades, as this topic has turned out to tie together many of Dupré's (legitimate) obsessions about biological individuality, classification, and processes. In addition, I suggested that some of us—either via 'homology' or via 'convergence'—have reached conclusions that are very similar to Dupré's. All in all, the breadth and richness of Dupré's contribution to both philosophy of biology and conceptually and theoretically-informed biology are truly exceptional, and I bet that this a process that will still last very long.

Acknowledgements I would like to express here my most sincere thanks to John Dupré for inspiration, discussions, and twenty years of friendship, and more generally for being such a kind and generous person in the too often red in tooth and claw atmosphere of the academic world. I would also like to thank Sabina Leonelli for many years of friendship and building connections between Exeter and Bordeaux—and now between Munich, Exeter and Bordeaux! Adrian Currie, Chee Wong, Adam Toon, Stephan Güttinger, Celso Neto, Rose Trappes have been not just excellent scholars but also fantastic hosts at Egenis. I look forward to future interactions with John and all of them—as well as with the other contributors to the present volume. I would also like to thank Alexandre Guay for his collaboration on the concept of genidentity and interdisciplinary comparisons of individuality debates in physics and biology. This research is funded by the Gordon and Betty Moore Foundation through grant GBMF9021 to Thomas Pradeu.

References

Baedke, J., Fábregas-Tejeda, A., & Delgado, A. N. (2020). The holobiont concept before Margulis. *Journal of Experimental Zoology Part B: Molecular and Developmental Evolution, 334*(3), 149–155.

Bapteste, E., & Dupré, J. (2013). Towards a processual microbial ontology. *Biology & Philosophy, 28*(2), 379–404.

Bapteste, E., O'Malley, M. A., Beiko, R. G., et al. (2009). Prokaryotic evolution and the tree of life are two different things. *Biology Direct, 4*(1), 34.

Booth, A. (2014). Symbiosis, selection, and individuality. *Biology & Philosophy, 29*(5), 657–673.

Bouchard, F. (2009). Understanding colonial traits using Symbiosis research and ecosystem ecology. *Biological Theory, 4*(3), 240–246.

Bouchard, F. (2010). Symbiosis, lateral function transfer and the (many) saplings of life. *Biology & Philosophy, 25*(4), 623–641.

Bouchard, F., & Huneman, P. (2013). *From groups to individuals: Perspectives on biological associations and emerging individuality*. MIT Press.

Buss, L. W. (1987). *The evolution of individuality*. Princeton University Press.

Chiu, L., & Gilbert, S. F. (2015). The birth of the Holobiont: Multi-species birthing through mutual scaffolding and niche construction. *Biosemiotics, 8*(2), 191–210.

Clarke, E. (2011). The problem of biological individuality. *Biological Theory, 5*(4), 312–325.

Dupré, J. (1993). *The disorder of things: Metaphysical foundations of the disunity of science*. Harvard University Press.

Dupré, J. (2010). The polygenomic organism. *The Sociological Review, 58*, 19–31.

Dupré, J. (2012). Processes of life: Essays in the philosophy of biology. Oxford University Press. Retrieved November 11, 2015, from http://www.oxfordscholarship.com/view/10.1093/acprof:oso/9780199691982.001.0001/acprof-9780199691982

Dupré, J., & O'Malley, M. (2009). Varieties of living things: Life at the intersection of lineage and metabolism. *Philosophy & Theory in Biology, 1*.

Dupré, J., & O'Malley, M. A. (2007). Metagenomics and biological ontology. *Studies in History and Philosophy of Biological and Biomedical Sciences, 38*(4), 834–846.

Frandsen, T. F. (2017). Are predatory journals undermining the credibility of science? *A bibliometric analysis of citers. Scientometrics, 113*(3), 1513–1528.

Gilbert, S. F. (2002). The genome in its ecological context: Philosophical perspectives on interspecies epigenesis. *Annals of the New York Academy of Sciences, 981*, 202–218.

Gilbert, S. F. (2014). Symbiosis as the way of eukaryotic life: The dependent co-origination of the body. *Journal of Biosciences, 39*(2), 201–209.

Godfrey-Smith, P. (2013). Darwinian individuals. In F. Bouchard & P. Huneman (Eds.), *From groups to individuals: Evolution and emerging individuality*, pp. 17–36. MIT Press.

Godfrey-Smith, P. (2014). *Philosophy of biology*, Princeton University Press.

Guay, A., & Pradeu, T. (2016). To be continued: The Genidentity of physical and biological processes. In A. Guay & T. Pradeu (Eds.), *Individuals across the sciences*, pp. 317–347. Oxford University Press.

Haber, M. H. (2016). The individuality thesis (3 ways). *Biology & Philosophy, 31*(6), 913–930.

Haber, M. H. (2025, February 3). Biology's Einstein moment: Specifying lineal frames of reference and rejecting absolute biological history. *Biological Theory. Epub ahead of print*. https://doi.org/10.1007/s13752-024-00475-3

Hooks, K. B., & O'Malley, M. A. (2017). *Dysbiosis and Its Discontents. mBio, 8*(5), e01492–e01417.

Hull, D. L. (1978). A matter of individuality. *Philosophy of Science, 45*(3), 335–360.

Hull, D. L. (1980). Individuality and selection. *Annual Review of Ecology and Systematics, 11*(1), 311–332.

Hull, D. L. (1992). Individual. In E. F. Keller & E. A. Lloyd (Eds.), *Keywords in evolutionary biology* (pp. 181–187). Harvard University Press.

Kelley, D. (2024). The conceptual ecology of "Dysbiosis". *Biological Theory, 19*(4), 221–223.

Khelfaoui, M., Gingras, Y., Lemoine, M., et al. (2021). The visibility of philosophy of science in the sciences. *1980–2018. Synthese. Epub ahead of print*, (19 February 2021). https://doi.org/10.1007/s11229-021-03067-x

Lewin, K. (1922). *Der Begriff Der Genese in Physik, Biologie Und Entwicklungsgeschichte: Eine Untersuchung Zur Vergleichenden Wissenschaftslehre*. Springer.

Lloyd, E. A., & Wade, M. J. (2019). Criteria for Holobionts from community genetics. *Biological Theory, 14*(3), 151–170.

Maynard, S. J, & Szathmáry, E. (1995). The major transitions in evolution. Oxford, WHFreeman Spektrum.

Mayr, E. (1961). Cause and effect in biology: Kinds of causes, predictability, and teleology are viewed by a practicing biologist. *Science, 134*(3489), 1501–1506.

McFall-Ngai, M. J. (2002). Unseen forces: The influence of bacteria on animal development. *Developmental Biology, 242*(1), 1–14.

Méthot, P.O., & Alizon, S. (2014). What is a pathogen? Toward a process view of host-parasite interactions. *Virulence, 5*(8), 775–785.

Michod, R. E. (1999). *Darwinian dynamics: Evolutionary transitions in fitness and individuality*. Princeton University Press.

Müller, M., Mentel, M., van Hellemond, J. J., Henze, K., Woehle, C., Gould, S. B.,... & Martin, W. F. (2012). Biochemistry and evolution of anaerobic energy metabolism in eukaryotes. Microbiology and Molecular Biology Reviews, 76(2), 444-495.

Nicholson, D. J., & Dupré, J. (Eds.). (2018). Everything flows: Towards a processual philosophy of biology. Oxford University Press.

O'Malley, M. A. (2024). The concept of balance in microbiome research. *BioEssays, 46*(8), 2400050.

O'Malley, M. A., & Dupré, J. (2005). Fundamental issues in systems biology. *BioEssays: News and Reviews in Molecular, Cellular and Developmental Biology, 27*(12), 1270–1276.

O'Malley, M. A., & Dupré, J. (2007a). Size doesn't matter: Towards a more inclusive philosophy of biology. *Biology & Philosophy, 22*(2), 155–191.

O'Malley, M. A., & Dupré, J. (2007b). Towards a philosophy of microbiology. *Studies in History and Philosophy of Biological and Biomedical Sciences, 38*(4), 775–779.

O'Malley, M. A., Martin, W., & Dupré, J. (2010). The tree of life: Introduction to an evolutionary debate. *Biology & Philosophy, 25*(4), 441–453.

Padovani, F. (2013). Genidentity and topology of time: Kurt Lewin and Hans Reichenbach. In: N. Milkov & V. Peckhaus (Eds.), *The Berlin group and the philosophy of logical empiricism*. Boston studies in the philosophy and history of science 273. Springer Netherlands, pp. 97–122. Retrieved December 2, 2015, from http://link.springer.com/chapter/10.1007/978-94-007-5485-0_5

Parke, E. C. (2021). Trivial, interesting, or overselling? The microbiome and "what it means to be human". *Bioscience, 71*(6), 658–663.
Parke, E. C., Calcott, B., & O'Malley, M. A. (2018). A cautionary note for claims about the microbiome's impact on the "self". *PLoS Biology, 16*(9), e2006654.
Powell A, O'Malley MA, Müller-Wille S, et al. (2007). Disciplinary baptisms: A comparison of the naming stories of genetics, molecular biology, genomics, and systems biology. History and Philosophy of the Life Sciences 29(1). Stazione Zoologica Anton Dohrn-Napoli: 5–32.
Pradeu, T. (2010). What is an organism? An immunological answer. *History and Philosophy of the Life Sciences, 32*, 247–268.
Pradeu, T. (2011). A mixed self: The role of Symbiosis in development. *Biological Theory, 6*(1), 80–88.
Pradeu, T. (2012). The limits of the self: Immunology and biological identity. Oxford University Press, Paperback 2018a.
Pradeu, T. (2016). Organisms or biological individuals? Combining physiological and evolutionary individuality. *Biology & Philosophy, 31*(6), 797–817.
Pradeu, T. (2017). Thirty years of Biology & Philosophy: Philosophy of which biology? *Biology & Philosophy, 32*(2), 149–167.
Pradeu, T. (2018). Genidentity and biological processes. In: D.J. Nicholson & J. Dupré (Eds.), Everything flows: Towards a Processual philosophy of biology.: Oxford University Press, pp. 96–112. Retrieved October 27, 2018, from http://www.oxfordscholarship.com/view/10.1093/oso/9780198779636.001.0001/oso-9780198779636-chapter-5
Pradeu, T. (2019). *Philosophy of immunology*. Cambridge University Press.
Pradeu, T., & Carosella, E. (2006). The self model and the conception of biological identity in immunology. *Biology and Philosophy, 21*(2), 235–252.
Pradeu, T., & Vivier, E. (2016). The discontinuity theory of immunity. *Science Immunology, 1*(1), aag0479–aag0479.
Pradeu, T., Jaeger, S., & Vivier, E. (2013). The speed of change: Towards a discontinuity theory of immunity? *Nature Reviews Immunology, 13*(10), 764–769.
Pradeu, T., Lemoine, M., Khelfaoui, M., et al. (2024). Philosophy in science: Can philosophers of science permeate through science and produce scientific knowledge? *The British Journal for the philosophy of science* 75(2). *The University of Chicago Press*, 375–416.
Queller, D. C., & Strassmann, J. E. (2016). Problems of multi-species organisms: Endosymbionts to holobionts. *Biology & Philosophy, 31*(6), 855–873.
Reichenbach, H. (1956). *The direction of time*. University of California Press.
Roughgarden, J., Gilbert, S. F., Rosenberg, E., et al. (2017). Holobionts as units of selection and a model of their population dynamics and evolution. *Biological Theory*, 1–22.
Skillings, D. (2016). Holobionts and the ecology of organisms: Multi-species communities or integrated individuals? *Biology & Philosophy, 31*(6), 875–892.
Spribille, T., Resl, P., Stanton, D. E., et al. (2022). Evolutionary biology of lichen symbioses. *New Phytologist, 234*(5), 1566–1582.
Suárez, J. (2018). The importance of symbiosis in philosophy of biology: An analysis of the current debate on biological individuality and its historical roots. *Symbiosis, 76*(2), 77–96.
Suárez, J. (2020). The stability of traits conception of the hologenome: An evolutionary account of holobiont individuality. *History and Philosophy of the Life Sciences, 42*(1), 11.
Suárez, J., & Stencel, A. (2020). A part-dependent account of biological individuality: Why Holobionts are individuals and ecosystems simultaneously. *Biological Reviews. Wiley Online Library. Epub ahead of print, 2020.*
Swiatczak, B., & Cohen, I. R. (2015). Gut feelings of safety: Tolerance to the microbiota mediated by innate immune receptors. *Microbiology and Immunology, 59*(10), 573–585.

Tauber, A. I. (1994). *The immune self: Theory or metaphor?* Cambridge University Press.
Triviño, V., & Suárez, J. (2020). Holobionts: Ecological communities, hybrids, or biological individuals? A metaphysical perspective on multispecies systems. *Studies in History and Philosophy of Science Part C: Studies in History and Philosophy of Biological and Biomedical Sciences, 101323.*

Chapter 8
The Disorder of Processes: The Case of Human Nature

Tim Lewens

Abstract This chapter reviews John Dupré's career-long interest in the topic of human nature, by isolating four themes that have characterised his contributions in this area. First, Dupré tends to oppose accounts of human nature that ground it in supposedly 'biological'—as opposed to 'cultural'—facts. Second, he has resisted reductive definitions of human nature (whether 'biological' or not), which take the form 'X is a part of human nature iff Y'. Third, he has had an ambiguous relationship with eliminativism about human nature. He sometimes suggests that 'human nature' is a notion we would be better off without, and yet he also indicates that eliminativism should be resisted in favour of the claim that human nature is uniquely plastic. Fourth (and related to the third theme), Dupré has expressed doubts about how helpful the concept of human nature is. I argue that the unruly nature of the processes that explain similarities and differences within our own species and others helps to explain the appeal of all four themes. In so doing, I aim to show some payoffs of Dupré's process-based approach to life, while also vindicating his general stance on human nature.

Keywords Cultural evolution · Eliminativism · Essentialism · Human nature · John Dupré · Process ontology · Social learning

8.1 A World of Processes

A sign on the side-gate of King's College in Cambridge announces, 'No bicycles, dogs or picnics allowed'. Some metaphysicians may think that this prohibition ranges over a highly disunified collection: dogs are organisms, bicycles are artefacts, picnics are events. But John Dupré (2012a) tells us that they are all *processes*, albeit processes with salient differences between them. They differ in the speed with which they are concluded. A picnic is usually over far more quickly than a typical dog, and even a

T. Lewens (✉)
University of Cambridge, Cambridge, UK
e-mail: tml1000@cam.ac.uk

S. Leonelli et al. (eds.), *Life Beyond Human Boundaries: John Dupré's Philosophy and the Interdisciplinary Study of the Life Sciences*,
https://doi.org/10.1007/978-3-032-26977-5_8

healthy dog often lasts for less time than a well-maintained bicycle. They also differ with respect to their stability over time, the proper explanation of this stability, and the integrity of their constituent elements. A bicycle owes its stability to the resilience of its parts; and when those parts break, an external agent must fix them if the bicycle is to continue in operation. A dog, on the other hand, is moderately stable for much of its duration despite the ephemeral nature of the cells that make it up; and it is often the dog itself that takes care of repairing parts that are broken. A picnic's physical location is likely to remain more tightly confined than that of any dog or any bicycle, but it might expand and contract markedly as individuals join. And a picnic may demonstrate very little stability. From time to time, it might involve music, dancing, or Scotch eggs; but the picnic as a whole need not come to an end—although its character may change significantly—when one of these constituents peters out.

This flexibility of Dupré's processual metaphysics can be seen either as a strength or as a weakness. On the one hand, it can encompass so many diverse things! On the other hand, if everything—bicycle, dog, picnic, inter alia—is to be understood as a 'process', then what is the use of the category? There is a threat that, in the interests of an informative taxonomy, one will simply have to re-introduce sub-categories of types of process. These sub-categories may end up re-establishing the more traditional distinctions metaphysicians have wanted to draw between, for example, things and events (e.g. Mellor, 1980). This chapter aims to show, via attention to Dupré's own work on human nature, one way in which process philosophy has value. Instead of attempting to justify an appeal to processes by pointing to their unity, I will instead make the case that the disorder characterising the motley collection of processes underlying stability in the human species helps to show what is wrong with 'biological' accounts of human nature, and what might be gained by dispensing with the notion of human nature altogether.

8.2 Dupré on Human Nature

Human nature has been a topic of interest to Dupré over the course of his career, both before and after his overt turn to a process philosophy. Human nature is, for example, a theme of his (1987) paper 'Human Kinds', and it remains a concern in his most recent (2025) book *Everyone Flows*. I have found eleven additional works that are substantially devoted to the discussion of human nature in the 36 year period between 'Human Kinds' and the 2023 Gifford Lectures that formed the basis of *Everyone Flows*.[1]

In this section, I highlight four major themes of this *oeuvre*: their further clarification and assessment is spread over the remainder of the chapter. Philosophy, too, is a process, and it is unsurprising that one can detect both change and stasis in Dupré's thinking on human nature.

Theme 1: Scepticism of 'biological' accounts of human nature

[1] Dupré (1991, 1998a, 1998b, 2001, 2002, 2003, 2004, 2010, 2012b, 2015, 2018).

It will not be a surprise to readers of Dupré's work that he has shown consistent opposition to the accounts of human nature that feature in many works of evolutionary psychology (e.g. Dupré, 2001); however, the vocabulary he has used to make this claim has changed over time. In the earliest piece I will mention in this chapter, he places stress on what is, in our species, the preeminent influence of culture compared with biological evolution: '...the major forces that have an impact on humans are cultural rather than biological...' (1987, p. 338). In later work, Dupré instead begins to express doubts about the very idea of separating biology from culture: 'Cultural and biological evolution are intricately and inseparably intertwined in the history of our species' (2001, p. 99). Even before Dupré's attention turned more fully to a processual metaphysics for biology, he defended an image of evolution—including human evolution—as nothing more than an aggregated set of developmental processes. Indeed, this has helped to justify his scepticism of accounts of human nature that ground it preferentially in 'biological', rather than 'social' or 'cultural' factors: 'what humans are like is itself a matter of the constant interaction in human development between internal, narrowly biological, factors, and external, generally social factors' (2003, p. 110).

Theme 2: No reductive analysis of the human nature concept.

Some of the most widely discussed philosophical works on human nature in recent years give what one might call a 'reductive' analysis of the human nature concept. In using the term 'reductive', I do not mean to imply purely micro-physical, or even neuroscientific, reduction. I simply point to accounts that try to give a moderately detailed characterisation of what human nature is, stated using a set of concepts thought to be more basic. Edouard Machery (2008, p. 326) defends a reductive analysis in this sense, when he proposes that human nature, is the set of 'properties that humans tend to possess as a result of the evolution of their species'. Machery's account is also a detailed one: he goes to some lengths to explain what is meant by the claim that humans possess a property 'as a result of the evolution of their species'. Having given such a reductive analysis, Machery says it is the job of the sciences to fill out the details of which traits are, and which are not, parts of human nature so understood.

The closest thing I have found to a reductive analysis of human nature from Dupré is his (2001, p. 95) claim that, 'There is no more to human nature...than the developmental cycles that currently constitute human life'. But Dupré's other remarks give good reasons to think that while this might work as an account of the facts on which claims about human nature rest, it cannot work as a good analysis of the concept of human nature. He points out that human nature, '...is surely intended to be something fundamental, something more significant than merely what some or most people happen to do' (2003, p. 109). In other words, the notion of human nature cannot pick out any or every developmental cycle that happens to make up a human life. Instead, it draws attention to those with some kind of salience. In an early piece that deals explicitly with the issue, Dupré (1991) follows Bernard Williams in remarking that, 'The nature of a species, including our own, centrally includes its ethology; and the most striking and fundamental feature of human ethology is its existence within culture' (130). Here we do not see a general account of what it means to attribute a 'nature' to a species, but we do see a specific comment about our own 'nature' that regards our cultural embeddedness as particularly striking.

Theme 3: Awkwardness regarding eliminativism about human nature.

As I indicated above, Dupré has tended to make various substantive assertions about human nature—that it is plastic, that it is embedded within culture—while declining to develop any more general analysis of the concept, even of a non-reductive kind. This absence may help to explain his ambiguous relationship with eliminativism about human nature. One way to motivate eliminativism combines a clear analysis of the human nature concept (e.g. the proposal that equates the concept of human nature with those intrinsic traits that make an individual a member of *Homo sapiens*), coupled to a sceptical evaluation of whether anything in nature answers to that description (e.g. the claim that species have no intrinsic essences). A different way to motivate eliminativism might involve a general account of the role of discourse about human nature, leading to an overall evaluation that the concept tends to do harm overall. In other words, one might adapt Haslanger's general style of questioning, asking 'what work we want these concepts to do for us; why do we need them at all?' (Haslanger, 2000, pp. 33-4).

Dupré's overall stance on the question of elimination is awkward. For example, he has written: '…my sympathies lie more with the deniers of human nature than its defenders; but I prefer to say not that there is no human nature, but that human nature is uniquely plastic' (2015, p. 27). A few years later his emphasis was different: 'There is much to be said for a flat-out denial that there is any such thing as human nature…I think it is safer to dispense with its [human nature's] use altogether' (2018, pp. 104–5). On the one hand, Dupré expresses support for a substantive claim about human nature; on the other hand, he flirts with the idea that human nature is a notion one would be better off without.

Theme 4: Doubt about the practical value of the notion of human nature

Perhaps the key to understanding Dupré's awkward attitude to eliminativism lies in his doubts about the usefulness of the concept. He has recently described human nature as 'a generally unhelpful concept' (2025, p. 79): so perhaps the concept is *occasionally* helpful so long as it is understood in a way that highlights the plasticity of individual human developmental trajectories, and yet *generally* unhelpful since usually it is understood in some more misleading or otherwise objectionable way.

I have said that Dupré's position lacks any detailed analysis of human nature, while making numerous comments about human nature; that it is ambivalent regarding eliminativism about the notion; and that it hedges on the helpfulness of the concept. In what follows I want to make a case that Dupré is entirely justified in adopting this rather uncomfortable stance. It is just what the underlying processes and practices associated with evolution demand. Sections 4, 5, 6 and 7 of this chapter will make a case for each of the four themes I have drawn attention to here. Before I explain why I uphold these views, I need to give some context for recent thinking about human nature within the philosophy of biology. So that is the job for sect. 3.

8.3 The Modern Quasi-Consensus on Human Nature

David Hull's influential writing on human nature was based on his view 'that particular species lack anything that might be termed an "essence"' (Hull, 1986, p. 10). When Hull moved on to ask, 'But why is it so important for the human species to have a nature?' (11), he implied a close linkage between the two themes of systematics and human nature. His implicit view was that the job of the notion of 'human nature'—like the job of any species's nature—is to pick out the essence of *Homo sapiens*. It is clear from Hull's discussion that he meant by this something—perhaps a single property, perhaps a collection of properties, perhaps a moderate proportion of a cluster of properties—that each individual organism must possess if it is to be a member of the species in question.

Hull's approach to taxonomy denied that species membership was determined in this way. This view has remained largely uncontroversial among philosophers and biologists. Moreover, Hull took it that the absence of any essence of *Homo sapiens*—understood as some intrinsic property possessed by all members of the species—had consequences for thinking about human rights. Instead of arguing for the legitimacy of some notion of human nature on the grounds that the foundation of human rights demands it, Hull claimed it was better to take the absence of a respectable notion of human nature as evidence that human rights need some other form of justification: 'Why must we all be essentially the same in order to have rights? Why cannot people who are essentially different nevertheless have the same rights? Until this question is answered, I remain suspicious of continued claims about the existence and importance of human nature' (Hull, 1986, p. 12; see Nathan, 2024 for discussion and dissent).

Hull's argument uses anti-essentialism about species to deny the existence of human nature. On his view, good biology tells us there is no essence of the human species; so universal human rights cannot be justified by reference to something that makes each individual a member of *Homo sapiens*. But Dupré—a consistent anti-essentialist about species—is right not to move too quickly from this view about biological taxonomy to eliminativism about human nature.

There is a significant historical pedigree for uses of the human nature concept that deny it any taxonomic role. When David Hume, for example, made a case for the need to understand human nature, he was doing nothing more than asserting the importance of human psychology as a foundation for other forms of knowledge (Lewens 2018). As he put it:

> It is evident, that all the sciences have a relation, greater or less, to human nature: and that however wide any of them may seem to run from it, they still return back by one passage or another. Even Mathematics, Natural Philosophy, and Natural Religion, are in some measure dependent on the science of MAN; since they lie under the cognizance of men, and are judged of by their powers and faculties. (Hume, 1978)

A knowledge of human nature, on this picture, is nothing more than a knowledge of the 'powers and faculties' of humans: it is not knowledge of a set of necessary and sufficient conditions for species membership.

While Hume's claims about 'human nature' are often informed by a comparison of the capacities of humans with those of other species, the result of that comparison is to conclude that human nature is substantially shared with other species. He notes that in 'forming our notion of human nature, we are apt to make a comparison between men and animals, the only creatures endowed with thought that fall under our senses' (Hume, 1995, p. 82). Even so, Hume's substantive conclusion following this comparison is that, in the 'whole sensitive creation … every thing is conducted by springs and principles, which are not peculiar to man, or any one species of animals' (Hume, 1978, p. 397). If human nature is shared with many other species, it cannot be what determines membership of the human species.

Let me recap. In spite of Hull's insistence that biological taxonomy does not rely on any notion of a species's essence, the legitimacy of the notion of human nature remains in play. That is because 'human nature' can be viewed in a less demanding Humean manner. In this chap. I will simply assume that Dupré, like Hull, is right to eschew the essentialist approach to taxonomy. In the following sect. I will make some sceptical remarks about Machery's reductive approach to human nature, arguing that Dupré is right to resist any analysis of the human nature concept that identifies it with a specifically biological nature. In the subsequent sect. I will go further, and use Ramsey's contrasting account to suggest that the function of the human nature concept tells against all reductive analyses (see also Nathan, 2024 for a related position).

8.4 For an Eclectic Approach to Human Nature; for Dupré's First Theme

Machery proffers a Humean analysis of human nature. Recall his claim that Human nature is the set of properties that 'humans tend to possess as a result of the evolution of their species' (2008, p. 326). This raises the question of what it means to attribute the presence of a property to an evolutionary process. In response, Machery has said this: 'Only those traits that are appropriate subjects of ultimate explanations that appeal to organic evolutionary processes are constitutive of human nature' (Machery, 2012, p. 477).

This formulation in terms of explanation is meant to acknowledge that all traits rely on both genes and enculturation for their development. Even so, says Machery (2018), the species-wide presence of only some traits is explained in a satisfactory way by appealing to what he here calls 'organic evolutionary processes'. It would be a mistake, according to Machery, to include the ability to recognize the face of Taylor Swift within 'human nature', even if her popularity has grown to such an extent that almost everyone now has this ability. Machery would agree that the development of this ability draws on generic perceptual capacities that may well have been favoured by natural selection acting in the distant past. Nonetheless, he will say, one cannot legitimately explain why individuals the world over are better able to recognize Taylor Swift than John Dupré by appealing to organic evolutionary processes—selection, mutation, migration, drift—that have somehow had the upshot that the former ability is far more prevalent than the latter. Instead, the explanation for why the recognition of Taylor Swift, rather than of John Dupré, is ubiquitous is of a different sort. It appeals to processes that act on different timescales, including aspects of global communication and marketing, asymmetries of power between the music industry and academia, and so forth.

It would also be a mistake, according to Machery, to attempt to exclude the possession of trichromatic vision from human nature, by arguing that trichromacy is only present in large proportions of adults because of elaborate networks of social support that allow babies to be born and raised to trichromatic adulthood in the first place.

His point is that while a 'proximate' explanation for the development of trichromacy needs to appeal to a wide combination of inputs, an 'ultimate' explanation for how trichromacy became established in our lineage is narrower, perhaps referring to cycles of natural selection acting on individuals whose reproductive fitness was enhanced by their ability to detect ripe fruits against dappled backgrounds (Mollon, 1989).

Examples of these sorts—keep trichromacy in, keep the recognition of Taylor Swift's face out—help to elucidate the intention that lies behind Machery's effort to draw a boundary around human nature. But his reliance on 'ultimate explanations that appeal to organic evolutionary processes' to delineate this boundary leaves open the question of precisely what an 'organic evolutionary process' is. It is here that Dupré's own processual approach to evolution has bite.

Dupré considers evolutionary history to be constituted by nothing more than the chaining together of a series of individual developmental processes from origination to death. On Dupré's view, any explanation of large-scale change and stasis in the human lineage appeals to organic evolutionary processes, because all such processes—including social learning—have influence on organisms.

Now, one might reasonably worry that the problem here lies with Dupré, and not with Machery. David Lewis (1996) thinks of 'Moorean facts' as those propositions we know better than the premises of any philosophical argument to the contrary. One might contend that it is a Moorean fact that the recent explosion in the global ability to recognize Taylor Swift is not an evolutionary process, while the far less recent origination of trichromacy in our ancestors is. But genuine problems arise when one attempts to find a rationale for how to rule some such processes in, while excluding others. The appeal to a process philosophy earns its keep here because the relevant processes that combine to explain why our lineage changes, and why it stays the same, are so unruly. The worry I expressed in the opening paragraphs of this chapter drew attention to the troubling lack of unity in the diverse collection Dupré recognizes as 'processes'; but rather than deny that diversity, here it is instructive to take stock of it, because of the difficulty it creates in trying to rule on which processes should be regarded as bone-fide 'evolutionary' ones.

To give a flavour of the worries one should have here, consider the suggestion I implicitly adopted above when giving an intuitive presentation of Machery's position. One might suggest that 'organic evolutionary processes' are those processes treated in mainstream evolutionary models of the modern synthesis; processes such as selection, mutation, migration and drift. There are problems with this suggestion. Cavalli-Sforza and Feldman (1981) drew attention to the fact that learned traits can affect Darwinian fitness, giving rise to the possibility that the standard form of natural selection—whereby populations change by virtue of the different reproductive outputs of the organisms making them up—can be mediated by social learning (see Lewens, 2024). So this way of trying to appeal to the practice of evolutionary science does not do what Machery wants it to, which is to exclude traits whose 'ultimate' explanations appeal to social learning from the scope of human nature.

One might try a different approach, namely to assert that 'organic evolutionary processes' are those mediated by genetic variation. The problem with this suggestion is that it seems to rule out, by unprincipled fiat, the proposal that when changes in organic lineages are mediated by the inheritance of epigenetic variation they, too, are 'evolutionary'. I do not mean to imply that the evolutionary role of these epigenetic processes is uncontroversial: on the contrary, there are significant discussions about the degree to which epigenetic variation is reliably inherited across generations in different taxa, and also about the extent to which evolutionary relevance should be a simple function of fidelity in inheritance across multiple generations. However, it is because such questions are live ones that the evolutionary significance of such processes cannot be settled simply by announcing that the very meaning of 'evolution' is tied to genetic variation (Lala et al. 2024, Lewens, 2025).

Finally, it is important to note Dupré's own claim that evolutionarily relevant processes are not simply those acting on what is sometimes called 'vertical' inheritance. Even when inheritance is 'vertical', in the sense that it passes from parent to offspring, it need not be mediated by the germline. Lateral gene transfer, for example, has been shown to be important for microbial evolution, with increasing evidence indicating its significance for an understanding of change (including adaptive change) in multicellular taxa. Symbionts, too, can be a significant vehicle of inheritance, but they need not be passed from parent to offspring through the germline (see Dupré, 2012b).

So, there are good reasons to think that evolutionary change can include vertical inheritance acting on socially-learned variation. There are also good reasons to think that horizontal forms of genetic transmission, as well as diverse non-genetic inheritance processes, contribute to evolutionary change. Putting these considerations together, it seems arbitrary to exclude horizontal transmission mediated by social learning from processes properly regarded as 'evolutionary'. Such an exclusion seems even harder to justify when one notes the relevance of social learning—even when it has a 'horizontal' or 'oblique' form—for the traditional concerns of evolutionary theory such as speciation and adaptation (Lewens, 2012). Social learning can result in restricted reproductive interactions between different populations of the same species when (for example) they acquire different habitat preferences (Lala et al. 2024). And social learning can underpin the acquisition of adaptation. Cecilia Heyes (2018), for example, has made the case that the capacity to imitate is widely shared among humans, but its development and evolution should be explained by reference to learning in a multiplex manner.

I have not yet made any case that it is impossible to distinguish these different evolutionary processes, or that there is no interest in trying to untangle them. The evolution of lactose tolerance, for example, does not involve one homogeneous process, whereby both dairying and the ability to produce lactase have emerged together. On the contrary, it matters that dairying has spread through comparatively fast cycles of cultural reproduction, and that these cycles have interacted with slower cycles of natural selection acting to favour genetic variants that promote lactase persistence (Lewens, 2017). The point of the preceding discussion, instead, is to establish that the network of differences and commonalities that run between these different processes do not justify any decision to deny some of them the status of 'evolutionary' processes. There is, therefore, no good reason for an account of human nature that rests solely on 'evolutionary' processes to exclude them.

Even if these processes can be untangled, Dupré is right, it seems to me, to draw attention (even if only in passing) to the existence of hybrid processes that lie at the interface of these different forms of evolution. Laland and Brown (2018), for example, suggest that learned ways of acquiring, processing and storing food can have knock-on impacts on subsequent cognitive development made possible by a constructed environment, itself characterized by a richer and more reliable supply of nutrition. The resulting augmented cognitive capacities, in turn, enable the learning of these important ways of acquiring, processing, and storing food. In schematic terms, this is a case where behaviour X in organisms in generation 1 collectively modifies the shared environment of organisms in generation 2, such that organisms in generation 2 reliably develop to acquire trait Y, which in turn allows them to perform X. This form of transmission, underpinned by collective niche construction, is not a simple matter of the inheritance of learned behaviours. In this scenario, individuals do not learn how to acquire enhanced cognitive capacities. Instead, these capacities' reliable development depends on the provision of a suitably structured environment rich in nutritional resources. However, the maintenance of that environment relies on a set of learned skills, whose own acquisition is in turn contingent on a set of enhanced cognitive capacities. This is not a straightforward case of cultural transmission of cognitive capacities; but it is a case whereby the reliable acquisition of these capacities depends on a form of 'scaffolded' development made possible by learning. Once again, I see no reason to deny these processes the status of 'evolutionary', given their salient similarities with other forms of inheritance based on niche construction.

One could respond, all the same, with the suggestion that even if processes of cultural inheritance are genuinely 'evolutionary', and even if there exist hybrid processes that stand at the juncture of cultural and more obviously 'organic' forms of evolution, it remains the case that the term 'human nature' should not be attributed to traits whose presence owes itself solely to those cultural evolutionary processes. On what grounds might such an exclusion be justified? Maria Kronfeldner has offered a rationale: 'That there is a separate cultural evolution allows delineating different kinds of stability. One of these kinds of stability is exhibited by traits that are parts of human nature, and another kind is exhibited by traits that are part of human culture' (Kronfeldner, 2018, p. 156). My concern with Kronfeldner's conjecture is that, especially in the sorts of hybrid cases that Laland and Brown have highlighted, there is no salient form of stability carved off by distinguishing human culture from human nature. Dupré's process philosophy reminds investigators that generalisations true of (almost all) humans can be stabilised by different kinds of processes; moreover, the same set of undisciplined processes underpins generalisations that hold good at narrower and broader levels of magnification. Social learning can at least sometimes underpin species-wide generalisations that are highly robust (such as the widespread ability to imitate among all humans), just as natural selection acting on genetic variation can underpin generalisations whose scope is considerably narrower (such as the ability to digest lactose only in those populations with a significant history of dairying).

Writing elsewhere, Machery has explained that he uses the term, '"evolutionary information" as a cover term for the class of factors that explain the evolution of traits… As the nature of these factors is controversial among evolutionary biologists and philosophers of biology, I'll remain neutral about what constitutes evolutionary information' (Machery, 2018, p. 32). I have argued in this section that Machery is not so neutral in these comments, for his comments presuppose that whatever 'evolutionary information' is, it excludes individual and social learning. It is indeed uncontroversial to oppose an anything-goes account of evolutionary processes, or evolutionary information, according to which literally every token instance of individual or social learning counts as an evolutionary process. But the lesson to draw from this is not that social learning can never constitute an evolutionary process. The lesson, instead, is that 'evolutionary' processes pick out systematic effects across populations, as opposed to picking out idiosyncratic effects that only concern specific individuals. It is reasonable to deny that John Dupré's learning how to play the viola da gamba is an evolutionary process. But the reason for this exclusion is not that social learning cannot be a bona-fide evolutionary process. It is, rather, that this token instance of social learning does not have enough significance at the level of any population. For exactly the same reason, it is not appropriate to regard every token instance of migration (from Stanford to Exeter, say), every token genetic mutation, and every instance where an organism has a baby, as an evolutionary process either; but that does not mean that migration, mutation and selection are not properly regarded as evolutionary processes.

8.5 Against Reductive Analysis; for Dupré's Second Theme

The previous section showed that there is no good justification from within the biological sciences for excluding socially learned traits from the scope of human nature. That leaves open a rather different argument for this exclusion. One might argue that it is simply part of the entrenched meaning of the term 'human nature' that a trait cannot be part of 'human nature' if it owes itself to forms of enculturation. If someone suggests that human culture can be the grounds for some elements of human nature, one might suggest such a person simply does not know what 'human nature' means in everyday language.

This effort to build an argument from accepted use can be rebutted. Consider Hume on justice. Roughly speaking, his view is that humans collectively institute conventions—such as respect for private property—because of the advantages these conventions bring to animals like us that live in social groups. Even so, education teaches respect for these conventions, and this leads to an altered set of underlying motivational dispositions. Feelings of justice take on such 'firmness and solidity, that they may fall little short of those principles which are the most essential to our natures, and the most deeply radicated in our internal constitution' (Hume, 1978, p. 501). In other words, 'human nature' names that which is 'firm and solid', or that which has deep roots in our 'internal constitution', regardless of what processes—social convention, education, or otherwise—explain such traits' entrenchment.

For a very different example of use, consider Henrich et al.'s (2010) exposé of the over-reliance in psychological research on data drawn from what they call WEIRD—Western, Educated, Industrialised, Rich and Democratic—research subjects. They bemoan the fact that '…the available database [of psychological studies] does not reflect the full breadth of human diversity. Rather, we have largely been studying the nature of WEIRD people, a certainly narrow and potentially peculiar subpopulation' (2010, p. 63). Here, the 'nature' of WEIRD people refers to the moderately robust traits of WEIRD people, even if that robustness is due to forms of cultural entrenchment. Similarly, when they refer to the 'unusually individualistic nature of Americans' (74), it would entirely undermine the intention of their article to suggest this is a claim about what Americans are like 'by nature' rather than 'by culture'. The remark simply makes a claim about a robust feature of Americans. 'Human nature' is simply a more general category than WEIRD nature, which in turn is more general than American nature. And if Americans, or WEIRD people, can have moderately stable 'natures' that are not meant to be contrasted with their supposedly more ephemeral 'cultures', then there is no conceptual mistake is suggesting that some elements of human nature might also owe their stability to entrenched features of human culture.

Arguments from use are rarely conclusive, because usage is so variable. Nonetheless, I have produced some evidence for thinking that it is not built into the accepted meaning of human nature that it has to be distinguished from human culture. Might one then defend an alternative reductive analysis of the human nature concept, which is far more liberal than Machery's in what it includes within the scope of human nature? In the remainder of this sect. I will cast doubt on the prospects for such a reductive account, while retaining the idea that 'human nature' can encompass human culture. I do so via a discussion of Grant Ramsey's (2018, 2023) maximally inclusive Humean analysis of human nature, which he calls the Life-History Trait Cluster account, or LTC account. It decouples an account of human nature from an account of species membership, while placing stress on the plasticity of developmental processes.

Ramsey says that human nature is 'the pattern of trait clusters within the totality of extant human possible life histories' (2023, p. 21). I will say a little more about what this formulation means in a moment. First, let me explain Ramsey's motivation for his position. Ramsey explains his account by drawing an analogy between human nature and the nature of a baseball:

> …the nature of a baseball lies in how it performs in various circumstances. When hit with a bat with a particular angle and velocity, how much is it deformed and how far will it travel? Baseballs do not act outside of their nature, but this does not mean that they do not have a nature, or that they have an overly permissive nature. Baseballs act in very specific, reliable ways given their particular environmental inputs, and it is their nature to do so. (2018: 46)

This way of phrasing things may create an expectation in the reader—at least the reader familiar with metaphysical debates about properties—for what Ramsey's account will go on to say about human nature. It encourages one to think that the nature of a baseball consists in its various dispositional properties. One spells out what these dispositions are by specifying how the baseball would react under various environmental triggers. These dispositions place constraints on what one can, and cannot, rightly say (in conditional terms) about a baseball's nature: it is in its nature to briefly deform when struck hard with the typical force of a batter; it is not in its nature to vaporize under those same circumstances. Both claims are simply ways of describing the baseball's elasticity.

Consonant with this, Ramsey moves on to make analogous claims about the nature of both Socrates as an individual, and about humans more generally: 'Socrates's death is explained by his consumption of hemlock. It is our nature to die from its consumption. We could go deeper than this, pointing out that hemlock contains coniine, and that coniine consumption leads to muscle paralysis, that among the muscles affected are those involved in respiration, and that with all but the shortest absence of respiration, humans die' (2023, p. 42). So just as a baseball has various dispositions, so does Socrates. In both cases, these dispositions are well described using conditionals: both Socrates, and humans in general, are such that if they imbibe hemlock, they die. One might think Ramsey says that human nature consists in what humans are *generally* like: in other words, human nature picks out widely shared human dispositions (whether those dispositions are manifested or not). These dispositions are picked out using conditionals, which in turn associate patterns of triggering conditions and responses. Hence, they are trait 'clusters', because they correspond to the properties that are moderately tightly 'clustered' in all humans.

In the end, this is not the account that Ramsey settles on. Ramsey denies that the term 'human nature' needs to pick out properties that nearly all humans possess, and I agree with him about that. Over a century ago, E. B. Titchener argued, in a vigorous paper responding to Rivers's (1901) work on Murray Islanders' perceptual capacities, that 'human nature is much the same the world over' (Titchener, 1916, p. 206). This claim only makes sense if it is also coherent to suggest that human nature might, in fact, have had features that vary from one part of the world to another; what is more, in this passage Titchener's primary point seems to be that humans vary in similar ways the world over, not that they are of a single standard design wherever they are found.

I also agree with Ramsey, for reasons I explained in the last section, that human nature can encompass human culture. I part with Ramsey's analysis in two respects. First, I suggest that Ramsey's account is equivalent to the claim that human nature names all the dispositions humans have; the vocabulary of patterns and conditionals is a way to describe those dispositions. Second, I place more emphasis than Ramsey does on the use of the term 'human nature' as a way to draw attention to important truths about humans.

The second amendment follows from the first. Ramsey includes all 'patterns' within the scope of human nature: that is to say, he includes all conditionals that one might assert about humans, regardless of whether they succeed in picking out widely shared dispositions or not. Part of the 'pattern' of human nature includes those conditionals that are true of only a small fraction of members of the species, because their truth depends on the instantiation of very rare dispositions. It is true of all humans that if they consume hemlock, then they die. It is only true of some humans that if they consume lactose, then they become sick. It is true of some other humans that if they consume lactose, then they do not become sick. And it is true of some humans that if they are exposed to a rich educational environment, they grow to love *The Disorder of Things*. These are all elements of 'the pattern of trait clusters within the totality of extant human possible life histories', hence they are all aspects of human nature.

Ramsey's view, as I understand it, is that this extreme permissiveness is a feature and not a bug. To give a full account of human nature is simply to offer an account (via a specification of a huge collection of conditional statements) of all the dispositions that can be attributed to humans. In this respect, an account of human nature is just like an account of the nature of an individual baseball after all, because there are no dispositions possessed by a baseball that fall outside its 'nature'. It is within the nature of an individual baseball that if one were to look at it, one would notice a discoloration on a tiny fraction of its surface. It is within the nature of humans in general that if they were to be surveyed, one would find that a tiny fraction of them can play the viola da gamba.

In fairness to Ramsey, he does acknowledge the desire for some kind of prioritisation in the patterns one picks out (2023, pp. 22–23). His preferred way of achieving this is via the notions of 'pervasiveness' and 'robustness' (Ramsey, 2013). The first quantifies how often the antecedent of a conditional is satisfied, the second quantifies how reliable the link is between antecedent and consequent. Compare the two conditionals, 'If someone eats regularly through their childhood, they will grow to at least 160 cm in height' and 'If someone is exposed to a disciplined classical music training through their childhood, they will grow to be able to sight-read proficiently'. The first conditional scores much higher on 'pervasiveness' than the second, for the simple reason that hardly anyone is exposed to a disciplined classical music training throughout their childhood, while much higher proportions of people eat regularly through their childhood. The second conditional scores much higher on 'robustness' than the first, because while a regular diet by no means guarantees that one will reach a height of 160 cm, a disciplined classical music training gets far closer to ensuring sight-reading proficiency. Ramsey does not say that only conditionals that score highly on both counts are elements of human nature; however, his account does allow him to explain the excitement attached to a claim like 'if someone is exposed to regular linguistic stimuli through early life, they will learn a language possessing some richly specifiable set of Chomskyan features'. A conditional like this scores highly on pervasiveness and robustness. It also has a further attractive aspect that Ramsey does not discuss, namely the informativeness of the consequent.

The term 'human nature' can be used in a highly flexible way, to draw attention to truths about humans that are important as one considers humans in general. Such importance need not be grounded only in Ramsey's highlighted features of pervasiveness and robustness. I have already pointed to the additional importance of informativeness. Hume (1978) suggests a different form of importance when he asks, optimistically, '...may we not hope, that philosophy, if cultivated with care, and encouraged by the attention of the public, may carry its researches still farther, and discover, at least in some degree, the secret springs and principles, by which the human mind is actuated in its operations?....What though these reasonings concerning human nature seem abstract, and of difficult comprehension?' So 'human nature', for Hume, picks out that which is significant because it is secret, or hidden. Machery's own account also aims to delineate a set of properties humans possess that have particular importance; his mistake, I have suggested, is in thinking that the right form of importance cannot be found in traits whose origin explanations lie with cycles of social learning.

What makes such truths 'important' is exceptionally open-ended, and may reflect such considerations as: (i) richly characterised properties of humans that it is valuable for anyone to be familiar with, because such familiarity is likely to assist in understanding more or less anyone whom one might encounter; (ii) striking, even if unusual, features of humans whose significance lies in the fact that they are rarely, if ever, seen in members of other species; (iii) features of human populations as a whole, which may reflect variegated distributions of properties within populations, rather than properties shared by all individuals. Putting these considerations together reinforces my scepticism of the need for what I have called a reductive analysis of human nature. Human nature consists in properties that are important for understanding humans in general, and analysis cannot give further specification than that without excluding too much.

8.6 Against Eliminativism about Human Nature; for Dupré's Third Theme

Scientific American is a periodical aimed at a wide audience of general readers interested in science. In 2012 it ran an article under the title 'Scientists probe human nature—and discover we are good, after all' (Ward, 2012). What followed was a summary of research indicating that automatic, unreflective interactions with respect to others tended to be cooperative, while slower, more rationally engaged actions, tended to be more self-regarding. That *Scientific American* article also ruminated, briefly, on whether 'human nature' can reasonably be used to describe humans' spontaneous motivations, or whether elements of 'human nature' also need to be inborn. It came to no conclusion on this matter. Tellingly, the more technical journal article that was summarized in this popular piece (which appeared in *Nature*) did not use the phrase 'human nature' at all, or indeed anything approximating to it. Instead, it simply made the case (based on experimental data) 'that intuition supports cooperation in social dilemmas, and that reflection can undermine these cooperative impulses' (Rand et al., 2012, p. 427).

This vignette fits remarkably well with Mameli's (2024) view that the function of the notion of human nature is to act as what he calls a 'channel of communication', mediating between claims in the sciences and a range of social and political aspirations that make those scientific claims salient. This is a case where the original scientific research, as presented to peers, was not flagged up using the 'human nature' label; only when that research was reported to a broader audience, in the context of the science journalist's wide-ranging allusions to the political projects of Hobbes and Rousseau, was the term 'human nature' used to signal the significance of Rand et al's findings.

My own view is very close indeed to Mameli's. The only point of difference is that I accord a broader function to the human nature concept. It can be used as a more general flag of significance, including when scientists are highlighting results whose importance is relevant for other scientific projects. Not all cases are like the *Scientific American* example: sometimes the term 'human nature' features in more technical articles intended for specialists; but when it does, it is often in the sections that frame the research, or which summarise its significance for other scientific readers. My conjecture is that the function of the notion of 'human nature' is to drawn attention to some set of well-confirmed claims about what humans are like, constrained only by their significance. Significance, in turn, is relative to the interests of the investigator.

Consider once again Henrich et al.'s (2010) article on the 'Weirdest People in the World'. This paper appeared in *Behavioral and Brain Sciences*: a respected academic journal, rather than a popular science magazine. Henrich et al. remark in the abstract of the piece that 'Behavioural scientists routinely publish broad claims about human psychology and behavior in the world's top journals based on samples drawn entirely from Western, Educated, Industrialised, Rich and Democratic (WEIRD) societies.... Overall, these empirical patterns suggest that we need to be less cavalier in addressing questions of *human* nature on the basis of data drawn from this particularly thin, and rather unusual, slice of humanity' (2010, p. 61, emphasis in original). The phrase 'human nature' is almost entirely absent from the remainder of this long article. It only reappears once, close to its conclusion, where the authors reflect that:

> Research partnerships with non-WEIRD institutions can be established to further the goal of expanding and diversifying the empirical base of the behavioral sciences. By supplying research leaves, adjusted expectations of student progress, special funding sources, and institutionalized relationships to populations outside the university as well as to non-WEIRD universities, these organizations can make an important contribution to building a more complete understanding of human nature. (2010, p. 82)

The authors are clear that something important about humans will be learned by undertaking these more wide-ranging studies, while remaining non-committal on how 'human nature' is being understood. In particular, their phrasing is neutral about whether 'human nature' picks out only those traits that are genuinely pan-cultural, or whether these more diverse studies will give a better evidence base for a variegated picture of human nature as something that differs from one cultural group to another. In the responses they give to peer-reviewed comments, they mention that their research does not 'imply an absence of a universal human nature' (114), but even this phrasing seems to leave open the coherence of a human nature that is not universal. The article can afford to be neutral on this question, because what is abundantly clear is that its goal is to establish that scientists have been too quick to infer universals based on a narrow evidence base. This conclusion can be secured and communicated regardless of whether 'human nature' itself is understood solely to pick out those features of the species that are found across all cultures, or whether it can also draw attention to the patterns of variation seen between cultures.

I take the preceding discussion to have vindicated almost all of the elements of Dupré's treatment of human nature. To recap, claims about human nature are nothing more than claims about what humans are like; such claims can be entirely unconstrained, except that they must draw attention to aspects of humans that have some kind of importance for an understanding of humans in general. This means Dupré is right not to offer any more detailed definition, or conceptual analysis, of the notion of human nature. It also means that, in spite of his scepticism about some specific claims for what human nature is, and about some detailed proposals for analyses of human nature, it is entirely warranted to point (as Dupré does) to the importance of understanding the plasticity of developmental processes, to the role of cultural embeddedness in explaining the origination and perpetuation of various adaptive capacities, and so forth. To summarise these things by announcing that 'human nature' is plastic, culturally embedded, and such like is a reasonable use of the term.

8.7 The Risks of 'Human Nature'; for Dupré's Fourth Theme

But what should one say about the final theme of Dupré's approach to human nature? While Dupré does not embrace eliminativism in a wholehearted way, he does express concerns about the value of the human nature concept, and inclines to the view that it may, after all, be a concept we would be better off without. In this final section, I argue that this is a coherent—and even an attractive—stance to adopt.

Here is my strategy. On the one hand, as I have already indicated, the function of the human nature concept is to draw attention to various important facts about what humans are like. There is nothing objectionable in a concept that does these things, even if some uses of the term 'human nature' have leant on spurious facts, and even if the importance accorded to some sets of facts rests on objectionable priorities. So these considerations all speak against eliminativism. Nonetheless, 'human nature' is not the only term one can reach for to assert the importance of various truths about humans: one can instead write of 'the human condition' (as Mameli mentions), or even simply 'how humans are'. Moreover, there are reasons for concern about some of the connotations that many seem to read into the term 'human nature'. If one approaches the topic from a more pragmatic perspective, the arguments for getting rid of the notion of human nature start to look much stronger.

Let me begin with what may be the least controversial part of this series of warnings about the human nature concept. Various lines of research have suggested that many humans are 'folk essentialists' (Linquist et al., 2011). In other words, people have a tendency to attribute hidden teleo-essences to individual organisms, which correspond to the type of organism they are. These are termed ‘teleo-essences’, because they make the individual a member of the biological kind it belongs to, while also directing the development of the outwardly visible properties characteristic of the kind. There are risks that in making claims about human nature one triggers and cements these forms of folk essentialism. Moreover, folk essentialism has the power to do harm. The problem lies in the fact that folk essentialism does not deny the existence of significant behavioural (and even cognitive) diversity; instead, it hypothesises that when these forms of diversity are observed, they are expressive of a shared underlying nature mis-firing, and failing to produce its proper outcome on various occasions. This means, in turn, that it can be very difficult to dislodge folk essentialist views in the face of apparent counter-evidence: folk essentialism is not dislodged by pointing to differences between individuals. It also means that forms of social learning are more prone to being categorised as distorting, or masking, appropriate universal developmental responses, instead of contributing to appropriate local developmental responses. When they present evidence for this feature of folk essentialism, Linquist et al. (2011, p. 445) suggest that scientists should be able to accommodate a ‘non-universalist, developmentally plastic view of human nature’: however, their work also lends weight to the proposal that, because of the risks presented by the folk essentialist notion of human nature, it would be better to describe these patterns of human difference using a different vocabulary altogether.

This proposal receives further support from a less obvious line of argument offered by Dupré. Recall that when Hume discusses human nature, he does not merely reserve the term for truths about humans that he regards as important. What he says is considerably more specific. Human nature is bound up with that which has ‘firmness and solidity’; the features that are the ‘most essential to our natures’ are also ‘the most deeply radicated in our internal constitution’ (Hume, 1978, p. 501). So, for Hume, elements of human nature are important because they are robust, and they are robust because they derive from features that lie within us; indeed, this corresponds to a widely-held idea that the political value of understanding human nature is that one thereby understands which features of humans one must treat as largely fixed constraints, compared with others that might be more malleable in the face of reforms (Sterelny, 2018).

Dupré has pointed out that ‘what humans are like is itself a matter of the constant interaction in human development between internal, narrowly biological, factors, and external, generally social factors’ (2003, p. 110). This comment draws attention to two very different explanations for why some set of traits might be highly robust, in the sense that they are found again and again in human populations with great reliability. The most obvious one—suggested by Hume’s phrase, ‘deeply radicated in our internal constitution’—is that the traits in question develop in the same way in spite of broadly varying developmental environments (they are, as Ariew, 1996 puts it, highly canalized). But there is a quite different explanation that should also be in play: perhaps the traits in question only develop in a very narrow range of suitable environments. They are nonetheless ubiquitous, simply because those environments are very reliably generated. A suitable environment might appear reliably because of the stabilizing actions of earlier organisms, or because of the stability of various physical processes. Dupré cautions against exploring only the first (canalization) option, warning against the temptations of the 'centrality of the endogenous'. This is the approach according to which 'strategies of investigation that start with the influence of the environment are seen as worthless and the only properly scientific way forward is investigation of the intrinsic tendency of the organism to emit behaviour of various sorts' (2001, p. 72).

Putting all of this together, there is another risk associated with the notion of human nature. So far in this chapter, I have largely been discussing human nature as a vehicle for drawing attention to important truths about humans; this may happen in tandem with some kind of explanatory story about how these traits come to be, or what their effects are. But sometimes the term 'human nature' is also used to gesture to a kind of explanation: in asking whether some pattern of behaviour is a consequence of 'human nature', one gives the impression that the pattern in question is explained by appeal to some kind of internal state of humans, and that it cannot be changed without intervening on that internal state. Note, however, that the claim that some feature of humans is important—and even the claim that it is important in the sense of developing robustly—allows that it might change quite easily if only the relevant developmental environment were altered. The reason for robust development may lie in a lack of collective will to alter the developmental niche, rather than any literal impossibility in doing so. So here is another reason to worry about the notion of 'human nature': it can encourage a kind of unwarranted conservatism, by making it easy to elide under one label two phenomena that should be kept apart. They are (i) the pervasiveness of a trait as a target of explanation, and (ii) the insensitivity of development to changes in environment as a potential source of explanation for that same pervasiveness.

Once again, it seems to me that a term like 'human nature' is more likely than closely related terms such as 'the human condition' to engender this elision. In writing of 'the human condition' one retains a focus on important and pervasive truths about humans, while making it harder to think that the very same 'human condition' need also encode information about highly canalized internal features of humans. So, to close, Dupré is right that there are many illuminating things one can say about human nature, while also drawing attention to the benefits that might come from dispensing with the notion altogether.

Acknowledgements I am grateful to the audience at the University of Exeter, where this work was first presented in April 2024 at the 'Understanding Life in a Changing Planet' meeting. I would also like to thank Sabina Leonelli for inviting me to that meeting. I am grateful to Harriet Fagerberg, Jules Macome, Arsham Nejad-Kourki, Philipp Spillmann, Cassandra Yang and Jolie Zhou for comments on a written draft of the piece. Most important of all, I am grateful to John Dupré, who has taught me a great deal, shown me great kindness, and demonstrated what it is possible for a philosopher of biology to achieve.

References

Ariew, A. (1996). Innateness and canalization. *Philosophy of Science, 63*(Proceedings), S19–S27.

Cavalli-Sforza, L., & Feldman, M. (1981). *Cultural transmission and evolution: A quantitative approach.* Princeton University Press.

Dupré, J. (1987). Human Kinds. In J. Dupré (Ed.), *The latest on the best.* MIT Press.

Dupré, J. (1991). Reflections on biology and culture. In J. Sheehan & M. Sosna (Eds.), *The boundaries of humanity: Humans, animals, machines* (pp. 125–131). University of California Press.

Dupré, J. (1998a). Normal people. *Social Research, 65*, 221–248.

Dupré, J. (1998b). Against reductionist explanations of human behaviour. *Aristotelian Society Supplementary, 72*, 153–172.

Dupré, J. (2001). *Human nature and the limits of science.* Oxford University Press.

Dupré, J. (2002). 'Darwin and human nature' in J. Dupré *Humans and Other Animals.* Oxford University Press.

Dupré, J. (2003). On human nature. *Human Affairs: Postdisciplinary Humanities and Social Sciences Quarterly, 2*, 109–122.

Dupré, J. (2004). Human kinds and biological kinds: Some similarities and differences. *Philosophy of Science, 71*, 892–900.

Dupré, J. (2010). How to be naturalistic without being simplistic in the study of human nature. In M. de Caro & D. Macarthur (Eds.), *Naturalism and normativity* (pp. 289–303). Columbia University Press.

Dupré, J. (2012a). *Processes of life.* Oxford University Press.

Dupré, J. (2012b) 'Causality and human nature in the social sciences' in J. Dupré *Processes of Life.* Oxford University Press.

Dupré, J. (2015). Pluralism and process in understanding human nature. *Rivista di Filosofia Neo-Scolastica, 1-2*, 15–27.

Dupré, J. (2018) 'Human nature: A process perspective' in Hannon and Lewens (eds.) Why we disagree about human nature. Oxford University Press.

Dupré, J. (2025). *Everyone flows.* Oxford University Press.

Haslanger, S. (2000). Gender and race: (what) are they? (what) do we want them to be? *Nous, 34*, 31–55.

Henrich, J., Heine, S., & Norenzayan, A. (2010). The weirdest people in the world? *Behavioral and Brain Sciences, 33*, 61–135.

Heyes, C. (2018). *Cognitive Gadgets.* Princeton University Press.

Hull, D. (1986). Human Nature. *Proceedings of the Biennial Meeting of the Philosophy of Science Association, 2*, 3–13.

Hume, D. (1978). *A treatise of human nature.* Clarendon Press.

Hume, D. (1995). *David Hume: Essays moral, political, and literary.* Liberty Press.

Kronfeldner, M. (2018). *What's left of human nature?* MIT Press.

Laland, K., & Brown, G. (2018). The social construction of human nature. In E. Hannon & T. Lewens (Eds.), *Why we disagree about human nature* (pp. 127–144). Oxford University Press.

Lewens, T. (2012). Human nature: The very idea. *Philosophy and Technology, 25*, 459–474.

Lewens, T. (2017). 'Human nature, human culture: The case of cultural evolution' *Interface. Focus, 7*, 20170018. https://doi.org/10.1098/rsfs.2017.0018

Lewens, T. (2018). The faces of human nature. In E. Hannon & T. Lewens (Eds.), *Why we disagree about human nature.* Oxford University Press.

Lewens, T. (2024). *Cultural selection.* Cambridge University Press.

Lewens, T. (2025). Development Developing. *Biology and Philosophy, 40*, 15. https://doi.org/10.1007/s10539-025-09985-1

Lewis, D. (1996). Elusive knowledge. *Australasian Journal of Philosophy, 74*, 549–567.

Linquist, S., Machery, E., Griffiths, P., & Stotz, K. (2011). Exploring the Folkbiological conception of human nature. *Philosophical Transactions of the Royal Society B, 366*, 444–453.

Machery, E. (2008). A plea for human nature. *Philosophical Psychology, 21*, 321–329.
Machery, E. (2012). Reconceptualizing human nature: Response to Lewens. *Philosophy and Technology, 25*, 475–478.
Machery, E. (2018). Doubling down on the Nomological notion of human nature. In E. Hannon & T. Lewens (Eds.), *Why we disagree about human nature.* Oxford University Press.
Mameli, M. (2024). *Why human nature matters.* Bloomsbury.
Mellor, D. H. (1980). On things and causes in Spacetime. *British Journal for the Philosophy of Science, 31*, 282–288.
Mollon, J. (1989). "Tho' she kneel'd in that place where they grew..." the uses and origins of primate colour vision. *Journal of Experimental Biology, 146*, 21–38.
Nathan, M. (2024). *The quest for human nature.* Oxford University Press.
Ramsey, G. (2013). Human nature in a post-essentialist world. *Philosophy of Science, 80*, 383–393.
Ramsey, G. (2018). Trait bin and trait cluster accounts of human nature. In E. Hannon & T. Lewens (Eds.), *Why we disagree about human nature.* Oxford University Press.
Ramsey, G. (2023). *Human Nature.* Cambridge University Press.
Rand, G., Greene, J., & Nowak, M. (2012). Spontaneous giving and calculated greed. *Nature, 489*, 427–430.
Rivers, W. H. R. (1901). Reports of the Cambridge anthropological expedition to Torres Straits: Physiology and psychology, Volume 2, Cambridge University Press.
Sterelny, K. (2018). Sceptical reflections on human nature. In E. Hannon & T. Lewens (Eds.), *Why we disagree about human nature* (pp. 108–126). Oxford University Press.
Titchener, E. B. (1916). On ethnological tests of sensation and perception with special reference to tests of color vision and tactile discrimination described in the reports of the Cambridge anthropological expedition to Torres Straits. *Proceedings of the American Philosophical Society, 55*, 204–236.
Ward, A. (2012). Scientists probe human nature—And discover we are good after all. *Scientific American.* https://www.scientificamerican.com/article/scientists-probe-human-nature-and-discover-we-are-good-after-all/#

Part III
Interdisciplinary Philosophy

Chapter 9
From Studies *of* to Studies *with* the Life Sciences

Jane Calvert

Abstract At the 2024 conference celebrating 20 years of Egenis, I was asked to speak on a panel on "Prospects for the Interdisciplinary Studies of the Life Sciences". It might seem strange, but the one word that stood out for me in this title was the word "of". This is because I have increasingly been wondering whether "of" is the right preposition for my engagements with the life sciences as a social scientist. In an attempt to grapple with this issue, in this chapter. I ask: what does it mean to do interdisciplinary studies *with* the life sciences, rather than *of* them? What are the benefits and challenges of working closely with scientists and engineers, not as objects of study but as collaborators? Could we extend this idea to think of collaborating with other organisms, and what would that look like? Is it possible to understand life itself as a collaboration (Dupré and O'Malley 2009)? And does this more expansive notion of studying *with* help us rethink our relationship to our changing planet?

9.1 Introduction

"With" is a word that has different meanings in different situations. Szymanski (2023) discusses the ways in which the word can be used in the context of baking bread: "I can bake bread with rye flour. I can bake bread with my mother. I can bake bread with my Danish dough whisk" (p.140). In these ways, "with" is being used to connect the author to a constituent, a companion or an implement. I am going to use "with" in this chapter in Szymanski's second sense, as a word that connects the subject to a companion—a companion that may or may not be human or living. For the purposes of this chapter, with is about togetherness, it is about collaboration.

In what follows, I start by explaining why I am uncomfortable with thinking in terms of interdisciplinary studies *of* the life sciences and why I want to move towards studies *with* the life sciences. In discussing the latter, my initial focus will

J. Calvert (✉)
University of Edinburgh, Edinburgh, UK
e-mail: jane.calvert@ed.ac.uk

S. Leonelli et al. (eds.), *Life Beyond Human Boundaries: John Dupré's Philosophy and the Interdisciplinary Study of the Life Sciences*,
https://doi.org/10.1007/978-3-032-26977-5_9

be on interdisciplinarity collaborations between people across disciplines. I go on to consider collaborations between humans and non-humans, showing how the life scientists I work with collaborate with the organisms they engineer. I then draw on Dupré and O'Malley's (2009) work that sees life itself as a collaboration, and I suggest that we can perhaps even extend the idea of collaboration to encompass the non-living. I end by arguing that opening ourselves up to being transformed by our objects of study will help us build new solidarities across disciplines, which are becoming increasingly important as we face planetary-scale changes.

But before I launch into the chapter proper, I will go back to November 2002, when I first arrived at Egenis—then the ESRC Centre for Genomics and Society—at the University of Exeter. I had recently completed my PhD at the University of Sussex, and I travelled from Brighton to Exeter in a hired van with all my possessions (which at the time mainly consisted of photocopied journal articles) to start my postdoc. I count myself as extremely lucky—and rather surprised—to have been hired by John Dupré, Barry Barnes and Steven Hughes, the three directors of the newly-formed centre. I had a fantastic time.

I was employed as a social scientist working in the field of science and technology studies (STS), not a philosopher. But something I came to value greatly was that there were no barriers between the conceptual/philosophical and the empirical/sociological work. Looking back over my five years at Egenis, it is telling that many of our posters and presentations emphasised both the philosophical *and* sociological dimensions of genomics in their titles and contents. These close connections between the conceptual and the empirical are consistent with John's naturalistic metaphysics, which he understands as being informed by the results of empirical enquiry—both natural and social scientific. In continuation with the ways of thinking that I absorbed at Egenis, in this chapter I draw on literature from the philosophy of biology and from the social sciences, without sharply distinguishing them. I also make use of my own empirical work.

As a social scientist, it is perhaps not surprising that I find empirical work conceptually informative and enriching. But here I do not just want to emphasise the importance of the findings of empirical work as a source of information, but the actual *doing* of empirical work. This does mark a difference between the social sciences and philosophy, because empirical work is something philosophers do not traditionally do (although it is common among philosophers of science in practice). Something I will address in this chapter is the difference this might make.

At Egenis I was given the freedom to explore my own ideas, but with continual intellectual inspiration, a great feeling of collegiality, and a real sense that academic research can be fun—which is something I have tried to take with me where ever I go. There was also a feeling of excitement at being at the start of something special. We took seriously the idea of being a centre for *genomics* in society, and delved into the question of what made genomes distinctive, drawing on the scientific, philosophical, historical and sociological literature on the topic. Steve Hughes, one of the three Directors, memorably defined the genome as the "space in which genes happen" (Hughes, 2005), which reverberated with all of us in the centre (and, although it did not strike me at the time, drew attention to the temporal and processual aspects of the genome, anticipating John's later work in process philosophy).

I started my postdoc at Exeter researching intellectual property in genomics. Heavily influenced by John's work on biological ontology, I asked how it was possible that genes and genomes could be understood as entities that could be owned and commodified (see for example Calvert, 2007). My work on genomics then led me to an interest in systems biology, a field that emerged in the early 2000s and which was, according to some accounts, a challenge to the reductionism of molecular biology. I worked on this topic with John and Maureen O'Malley, who was an incredibly important part of my time at Egenis and very influential on my thinking (see for example O'Malley et al., 2007). With this work on systems biology as a foundation, a visit to Egenis by the historian and philosopher Evelyn Fox Keller encouraged Maureen and I, along with two then Egenis PhD students Alex Powell and Jonathan Davies, to consider systems biology's sister discipline—synthetic biology. We wrote a paper outlining what we saw as the three key approaches to the field (published as O'Malley et al., 2008). My continued work on this topic coincided with my own move to the University of Edinburgh in 2007.

Becoming interested in synthetic biology was very consequential for me. Synthetic biology was different from anything I had studied before, because the scientists and engineers building the field invited me to become part of their work. Although the genomicists and systems biologists had been welcoming and happy to talk to me, they had not actively sought my involvement. One reason I think I was approached in this way was because I was at the time one of the few social scientists who had written on synthetic biology, thanks to the paper with Maureen O'Malley, mentioned above. My initial engagements with synthetic biology gained momentum, and led to me becoming enrolled, entangled and engulfed in the field. I have been so ever since. In fact, I have recently written a book trying to make sense of my experiences working in synthetic biology as an STS researcher over more than a decade (Calvert, 2024).

9.2 Three Ways of Relating to the Life Sciences

One thing my work in synthetic biology has particularly brought into focus for me is the different ways in which it is possible to relate to a field of study as an STS researcher. I co-authored a paper with nine other STS colleagues in which we explored the various roles we found ourselves playing in synthetic biology—from colleague to trophy wife to trickster (Balmer et al., 2015). In my book, I use rooms as an analytical lens, examining the affordances and constraints of different spaces for engaging with synthetic biologists—including the laboratory, the coffee room, the art studio and the policy room (Calvert, 2024). For the purposes of this chapter, I draw on a simple threefold categorisation of the different ways in which STS researchers can relate to their fields of study. This categorisation was developed in conversation with two STS colleagues I have worked closely with for several years: Robert Smith and Erika Szymanski. It is represented in the figure below (created by Robert Smith).

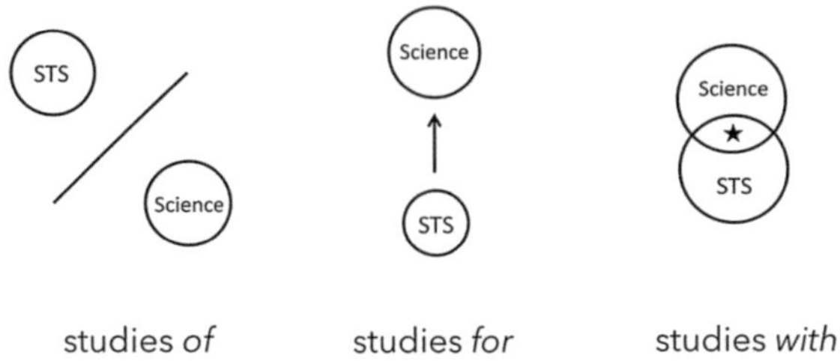

This categorisation starts with "of", the word that stood out for me in the title of the panel. We can do studies *of* science in an observational mode, which is perhaps the most familiar mode in which social scientists interact with scientist and engineers. Alternatively, we can do studies *for* science in a service mode. This might involve delivering bespoke teaching for natural science students or carrying out public engagement, for example. Finally, we can do studies *with* science in a collaborative mode. This is the mode that I find the most generative and interesting and is my focus here.

This threefold categorisation is a simplification of course, and it is not intended to be comprehensive. For example, in some circumstances we might want to study *against*. This is a mode that can involve disagreeing with people and challenging their views. Studying against is something I have sometimes described myself as doing when working in policy contexts (see Calvert, 2024). And there are of course other ways of relating to the life sciences that could be included in a larger categorisation scheme, but I will stick with the simple threefold one (of, for and with), because it encompasses the key features of my relationships with synthetic biology that I explore here.

The reason I became interested in thinking in terms of studying *with* was because I became increasingly dissatisfied with the traditional social scientific way of engaging with scientists: with studies *of*. Much STS work has criticised "the ocular, cerebral and objectifying gaze of the witness" (Atkinson et al., 2021, p.25), because of the separation it imposes between the observer and the observed. And I have found when working closely with scientists and engineers over long periods of time that it can feel extractive and ethically questionable to use them as sources of knowledge which one then goes away to write up for the benefit of oneself and one's peers. The synthetic biologists I have spent time with over the last 15 or so years have not been my objects of study. They have been my collaborators, my colleagues and my epistemic partners. I have not just studied and observed them, I have become part of their endeavours. Methodologically, ethically, and personally, it has become important for me to understand myself as studying *with* them.[1]

[1] Compare Adrian Haddock's chapter in this volume ("And the Wine is Bottled Poetry"), which argues that knowing what someone else thinks is fundamentally a matter of studying with them, rather than treating them as one's object of study.

It may be that I have reached this position because social scientific work engenders certain sensitivities and that it is because I am an empirical researcher that I have these qualms about observation. I find support for this suggestion from De Laet et al. (2021), STS researchers who emphasise the reflexive relations between social scientists and those they study, and argue that "knowledge, object and subject are co-produced in a complicated process of mutual shaping" (p.803). In other words, it is not necessarily clear who is studying who, or what will result from the interaction, instead this emerges through the encounter, through the process of studying with.

Since philosophers do not typically conduct empirical research, the qualms I am expressing about doing studies *of* may not resonate with them. In fact, they may (and in my experience often do) argue for the importance of autonomous reflection on the science from a disciplinary distance. I do not want to deny the value of this type of work in some circumstances. However, it is relevant to note that collaborative research is often not incentivised in the philosophy profession, and there is pressure to produce work primarily for one's philosophical peers, which may lead to a hesitancy amongst philosophers to invest in studying *with*.[2]

This issue aside, a philosopher may object that they do not have to concern themselves with the ways in which they relate to scientists and engineers, because they do not study them directly: they study the knowledge that they produce. But if we adopt a more interactional, relational view of knowledge production, which involves being open to being transformed by one's object of study, then philosophers are implicated in the process of studying with. I return to this point below.

9.3 Studying *with* as Interdisciplinary Collaboration

So what does studying with look like in practice? My first example concerns interactions between people from different disciplines—a form of studying with that fits under the broad heading of interdisciplinary collaboration.

One of my most rewarding experiences of this type of collaboration was a project I was involved in called Synthetic Aesthetics, which brought together synthetic biologists, artists, designers and social scientists (myself and my STS colleague Pablo Schyfter) to explore what it means to design with living things. I do not have space to discuss this project in detail here, but to give a snapshot of our activities, they included: asking how biology "computes" through shape and form; looking at synthetic biology from the humbling perspective of geological time; and making cheese from the bacteria that grow on human skin (see Ginsberg et al., 2014). This work amounted to a collaborative interrogation of synthetic biology that led us in directions none of us expected. But I think that the most important thing about this project was that what we did, we did together. We developed something at the intersection of our different disciplines that was unpredictable and emergent, and which would not have arisen without our interactions.

---

[2] I thank Sarah Green for making this point at the Egenis conference in 2024.

As noted above, this type of studying with can be understood as a form of interdisciplinary collaboration, but it is a type of interdisciplinarity "that promotes and benefits from the multiplicity and heterogeneity of ways of seeing that different disciplines offer" (Szerszynski & Galarraga, 2013, p.2817). It is not an understanding of interdisciplinarity in which each discipline adds their piece of knowledge to the same puzzle to complete a larger, unified picture. It is also not a managerial understanding of interdisciplinarity that is concerned with developing shared languages across disciplinary divides. It is a kind of interdisciplinarity that recognises and values diversity and pluralism—which is something that resonates strongly with John's work.

In fact, Szerszynski and Galarraga (2013) explicitly refer to John's work in advocating "a model of interdisciplinarity that involves heterogeneous methods and asymmetric contributions, underpinned by a model of the 'disunity of the sciences', epistemological pluralism (Dupré, 1993), and ontological multiplicity (eg, Mol, 2002)" (p.2817). Such a model requires humility and openness in recognising that there are radically different ways of understanding the world. This kind of interdisciplinarity is not necessarily easy but, as the Synthetic Aesthetics example showed, it can result in unexpected outcomes. By forcing everyone to become aware of (and challenge) their own assumptions it can expand the imaginations of everyone involved.

This is the ideal, of course. But studying *with* people has the danger that it can slip into studies *for*; that one can end up performing a service role for a more epistemically, financially or politically powerful group. And I am sometimes concerned that I have become an example of what could be called the "domestication of critique" in synthetic biology—that I have spent so long in the field I have become "tamed" by it and my critical capacities have become blunted. On some occasions, I worry that my presence has just enabled and facilitated what would have happened anyway, that I have taken up space that should have been filled by more radical and critical voices, such as those of NGOs and civil society organisations. And I have undoubtedly, albeit inadvertently, contributed to the construction of a politicised field (see Marris & Calvert, 2020). All this shows that studying *with* can potentially lead to compromise, complicity (Marcus, 1997), and contamination.

Contamination does not necessarily have to be regarded entirely negatively, however. In fact, I think it is key to studying *with*. Tsing (2015) argues that "Collaboration means working across difference, which leads to contamination" (p.28), and she usefully defines contamination as "transformation through encounter" (p.28). This gets to the crux of what I think is distinctive about the type of interdisciplinarity that involves studying *with*: it requires an openness to being transformed through encounters with those from different disciplinary perspectives. This can be enriching for all involved because such transformation can "enable unforeseen potentialities" (de Laet et al., 2021, p.812); it can show that things could be otherwise.

9.4 Collaborating with Other Organisms

That ends my brief exploration of studying with other people, but why stop at collaborations between people? I want to suggest that it is also possible to think about collaborations in a more extended sense—between humans and other organisms. I have come across these types of collaboration in my work on synthetic biology.

To provide some context to my discussion of collaborations between humans and other organisms, I need to say a little more about synthetic biology, a field that emerged in the early 2000s, driven by the attempt to apply engineering principles to biological systems (Endy, 2005). One of the early aims of the field was to build living systems from standardised interchangeable parts—parts that were often analogised to Lego bricks (Frow, 2013). But living systems tend to resist these attempts at standardisation. In fact, my investigation of synthetic biology has sensitised me to the "hopeful contingencies of biology" (Davies, 2011, p.439), which often overflow and exceed the constraints of engineering. This requires humility from the engineers, who have to learn to work *with* the biological systems they engineer and, to some extent, renege on their ambitions for control. In this way, I have seen the engineers transformed by the living systems they work with.

To give an example, one of the synthetic biology projects I have studied is the synthetic yeast project. This is an ambitious attempt to re-imagine and re-design the complete genome of the yeast species *Saccharomyces cerevisiae*—an important model organism, which is also commonly used in brewing and baking. The aim of this project is to make the yeast more predictable and amenable to human control (see Richardson et al., 2017).

Something that was particularly striking about the synthetic yeast project was the attachment that the scientists had to their organism. This was in contrast to the desire expressed by other synthetic biologists to become "organism agnostic" (Calvert & Szymanski, 2020)—to reach a situation where they did not require specific knowledge of the organism they happened to be working on, but could just apply general-purpose genetic tools across organisms. This is this why many synthetic biologists use the generic metaphor of the "chassis", drawn from mechanical engineering, when referring to their cells. But in the synthetic yeast project, the idea of organism agnosticism was completely undermined. In our ethnography of a synthetic yeast lab, Erika Szymanski and I observed a real appreciation for the yeast among the scientists and engineers. Rather than regarding it as a neutral chassis, they talked about how much affection they had for the yeast, how rewarding it was to work with, and how they liked the way it smelled. They even talked about how they wanted to preserve the "yeastiness" of the yeast in its synthetic form. They came across as totally enamoured by its organism-specific features (Calvert & Szymanski, 2020).

This takes us to the idea of collaboration, because in their enthusiasm for the yeast, the synthetic biologists sometimes explicitly described themselves as working *with* the yeast to achieve shared goals, through mechanisms like directed evolution and homologous recombination. They explained that because the work of building a new synthetic genome is shared between the synthetic biologist and the yeast, "we're working in collaboration with our organism, essentially" (Mitchell, 2017). The labour of the yeast was essential for the success of the research (Szymanski & Calvert, 2018).

Even given their enthusiasm and affection for the organism, it is necessary to point out that the scientists had an ambivalent—maybe even contradictory—relationship to the yeast, which they simultaneously regard as tools and partners, machines and collaborators. They killed a lot of yeast in the normal course of their experiments. Despite this ambiguity, I do think that this can be seen as an example of studying *with* that extends to incorporate collaboration between humans and non-humans.

9.5 Life as a Collaboration

Having moved from considering collaborations between people to considering collaborations between people and other organisms, I will now suggest that idea of collaboration can be extended even further to be understood as a central characteristic of life itself. Here I am drawing on, and inspired by, work by John and others.

Dupré and O'Malley (2009), for example, maintain that we should see life itself "as a continuum of variably structured collaborative systems" (p.1)—all the way from viruses to ecosystems with everything in between. Dupré and Guttinger (2016) similarly argue for "the fundamentally interconnected and collaborative nature of nature" (p.109). And Tsing (2015), in her study of the human and biological networks of matsutake mushrooms, shows that "staying alive—for every species— requires liveable collaborations…Without collaborations, we all die" (p.28).

In these understandings of the collaborative nature of life, *relations* become crucially important. Dupré and Nicholson (2018), in arguing that we should move away from a thing-based ontology of biology to a processual one, say that rather than regarding "relations as external to things", we should recognise that "organisms are fundamentally relational entities" (p.21), which "persist by virtue of the intricate webs of relations they maintain with one another" (p.21). Tsing also emphasises the importance of relations, and, as mentioned above, adds the idea of transformation, showing how matsutake "cannot live without transformative relations with other species" (p.40).

I see this emphasis on transformation as being consistent with John's process philosophy, because a "process depends on change for its occurrence" (Dupré & Nicholson, 2018, p.12), and transformation is a process of change. But I think that something that is particularly useful about the idea of *transformative* relations is that it connects back to the discussion above of how social scientists relate to their objects of study through a "process of mutual shaping" (de Laet et al., 2021, p.803). If we take this idea of mutual shaping, of transformation through encounter, in a more general sense, we could argue that things become through their relations (de Laet et al., 2021); that they are constituted in their relating (Szymanski, 2023).

This emphasis on collaborations and relations encourages us to see "withness" as something that is a central feature of life itself. But I think we could take a step yet further and argue that mutually transformative collaborations could extend to encompass those things we currently consider non-living. A version of this idea can be found in Dupré and O'Malley (2009). They view life as a continuum, which "leaves open the possibility that a variety of forms of organized matter – from chemical systems to ecosystems – might be usefully understood as living entities" (p.1). They argue when things are actively collaborating they are living, and this may extend to include chemical and physical systems, on some occasions. The lack of sharp distinctions between the living and the non-living also seems to follow from process philosophy, with its boundary-blurring emphasis on relations and interdependences.

9.6 Conclusions

In the panel session at the Egenis conference, we were tasked to "look ahead to the main opportunities and challenges for the interdisciplinary studies of the life sciences in our changing planet". In this chapter I have argued for a notion of studies *with* the life sciences that encompasses collaborations between people, between people and other living things, between living things, and between the living and the non-living. If collaboration is as pervasive as I have argued then this clearly has implications for our relationship to our changing planet, because we come to see ourselves as fundamentally interconnected with and dependent upon the living and non-living world.

I think such an emphasis on collaboration is important because it can help us build new solidarities across the natural sciences and the social sciences and humanities, which will becoming increasingly important as we face planetary-scale changes. But this requires a recognition amongst all of those involved (including philosophers) that we are implicated in the process of studying with. This in turn requires that we are all open to being transformed—contaminated even—through our encounters across disciplines. This requires both epistemic humility and continuous work.

I connect the idea of epistemic humility to John's research. I see it as consistent with his emphasis on epistemic pluralism and his recognition of the epistemic virtues of many fields of inquiry beyond those typically considered scientific (see Dupré, 1993). I think that openness to transformation through encounter can and does grow from this epistemic humility, although this is often not explicitly acknowledged. I would like to see a greater recognition across the board that knowledge is produced through interaction and in relation with others, and that opening oneself up to these others provides epistemic opportunities, which can be both conceptually enriching and hopeful. This might require more formal recognition of the value of collaboration within the structures of academia.

It is also important to note that collaboration requires continuous effort. Togetherness is not always easy, and constant work is required to maintain both biological (Dupré, 2024) and social forms (de Laet et al., 2021). To put it in another way, all collaborations deteriorate without care and maintenance. So being committed to collaborations is being committed to their care and maintenance, and to the investment of time and resources they require—through the creation of centres such as Egenis, for example.

Despite the challenges of studying with, I want to end by arguing for the importance of opening ourselves up to our objects of study, both human and non-human—to their hopeful contingencies and to being transformed through our encounters with them. And in terms of prospects, I hope to have shown how interdisciplinary studies *with* the life sciences—in all its various forms—can be a valuable way of working together as we move into an uncertain future.

Acknowledgements I would like to thank the organisers of the 2024 meeting 'Understanding Life in a Changing Planet: 20+2 Years of Egenis' and the participants for their excellent questions on the presentation on which this chapter was based. I am grateful to Adrian Haddock, Robert Smith and Erika Szymanski for their comments on a draft of this chapter. I acknowledge funding from the ESRC (ES/V002600/1) and the BBSRC (BB/Y008545/1).

References

Atkinson, S., Brives, C., Biederman, S., Cañada, J., Chartier, D., Davidson, A. C., & Evans, J. (2021). Introducing 'with microbes': From witnessing to withnessing. In C. Brives, M. Rest & S. Sariola (Eds.), *With Microbes* (pp. 17–40). Manchester: Mattering Press.

Balmer, A., Calvert, J., Marris, C., Molyneux-Hodgson, S., Frow, E., Kearnes, M., Bulpin, K., Schyfter, P., Mackenzie, A., & Martin, P. (2015). Taking roles in interdisciplinary collaborations: Reflections on working in post-ELSI spaces in the UK synthetic biology community. *Science and Technology Studies, 28*(3), 3–25.

Calvert, J. (2007). Patenting genomic objects: Genes, genomes, function and information. *Science as Culture, 16*(2), 207–223.

Calvert, J. (2024). *A place for science and technology studies: Observation, intervention and collaboration.* The MIT Press.

Calvert, J., & Szymanski, E. (2020). A feeling for the (micro)organism? Yeastiness, organism agnosticism and whole genome synthesis. *New Genetics and Society, 39*(4), 385–403.

Davies, G. (2011). Playing dice with mice: Building experimental futures in Singapore. *New Genetics and Society, 30*(4), 433–441.

de Laet, M., Driessen, A., & Vogel, E. (2021). Thinking with attachments: Appreciating a generative analytic. *Social Studies of Science, 51*(6), 799–819.

Dupré, J. (1993). *The disorder of things: Metaphysical foundations of the disunity of science.* Harvard University Press.

Dupré, J. (2024). The disunity of science and the Unity of the world presidential address, PSA 2022. *Philosophy of Science, 91*(5), 1043–1057.

Dupré, J., & Guttinger, S. (2016). Viruses as living processes. *Studies in History and Philosophy of Biological and Biomedical Sciences, 59*, 109–116.

Dupré, J., & Nicholson, D. (2018). A manifesto for a processual philosophy of biology. In D. Nicholson & J. Dupré (Eds.), *Everything flows. Towards a Processual philosophy of biology* (pp. 3–45). Oxford University Press.

Dupré, J., & O'Malley, M. (2009). Varieties of living things: Life at the intersection of lineage and metabolism. *Philosophy & Theory in Biology, 1*, 1–25.

Endy, D. (2005). Foundations for engineering biology. *Nature, 438*(24 November 2005), 449–453.

Frow, E. (2013). Making big promises come true? Articulating and realizing value in synthetic biology. *BioSocieties, 8*, 432–448.
Ginsberg, D., Calvert, J., Schyfter, P., Elfick, A., & Endy, D. (2014). *Synthetic aesthetics: Investigating synthetic biology's designs on nature.* The MIT Press.
Hughes, S. (2005). Navigating genomes: The space in which genes happen. *Tailoring Biotechnologies, 1*, 35–46.
Marcus, G. (1997). The uses of complicity in the changing mise-en-scene of anthropological field work. *Representations, 59*(13), 85–108.
Marris, C., & Calvert, J. (2020). Science and technology studies in policy: The UK synthetic biology roadmap. *Science, Technology, & Human Values, 45*(1), 34–61.
Mitchell, L. (2017, June 13th–16th). How we build and de-bug Sc2.0 chromosomes. Presentation at SB7: The seventh international meeting on synthetic Biology, National University of Singapore.
O'Malley, M., Calvert, J., & Dupré, J. (2007). The socioethical study of systems biology. *American Journal of Bioethics, 7*(4), 67–78.
O'Malley, M., Powell, A., Davies, J., & Calvert, J. (2008). Knowledge-making distinctions in synthetic biology. *BioEssays, 30*(1), 57–65.
Richardson, S., et al. (2017). Design of a synthetic yeast genome. *Science, 355*, 1040–1044.
Szerszynski, B., & Galarraga, M. (2013). Geoengineering knowledge: Interdisciplinarity and the shaping of climate engineering research. *Environment and Planning A: Economy and Space, 45*(12), 2817–2824.
Szymanski, E. (2023). Conversations with other-than-human creatures: Unpacking the ambiguity of "with" for multispecies rhetorics. *Rhetoric Society Quarterly, 53*(2), 138–152.
Szymanski, E., & Calvert, J. (2018). Designing with living systems in the synthetic yeast project. *Nature Communications, 9*, 2950. https://doi.org/10.1038/s41467-018-05332-z
Tsing, A. L. (2015). *The mushroom at the end of the world: On the possibility of life in capitalist ruins.* Princeton University Press.

Chapter 10
Why Do Philosophers Argue like They Do?

James Griesemer

Abstract This essay explores John Dupré's philosophical pluralism and processualism in and of the biological sciences and his pioneering theme of the disunity of science and nature. I trace some of the parallels in our thinking on these topics. A key goal is to consider consequences of his pluralism about biological classification in reflecting on his metaphysical stance toward process ontology and epistemology as the most suitable kind of philosophy for biology. I diagnose Dupré's philosophical strategy as one of taking models and modelers at face value for what they assume (metaphysically) and say (epistemically) about empirical "nature." I argue this is not always the best strategy for understanding the content of scientific practices, even if it is a good way of identifying value commitments of scientists and their scientific practices. I contrast "multi-perspectivism" with Dupré's philosophically pluralist disunity stance in the hope of honoring John's deep and profound contributions to philosophy of biology, philosophy of science, and philosophia in its wide sense as love of wisdom.

10.1 Introduction

I admire John Dupré's work in both style and substance, or perhaps I should say, in both activity and process. His work is never run-of-the-mill, nor has he pursued topics in a merely critical way: his criticism of scientific and philosophical business as usual has routinely opened new space for philosophical and scientific possibilities. Mere endorsement, though, makes it harder to discern, much less take a stand on, anything we may disagree about. In honor of his work, I will try to bring into relief some features of his views in terms of my own idiosyncratic reading, where we may differ in subtle but important respects or degrees. Because my views are idiosyncratic and occasionally unorthodox as well, I think the best way I can honor John and his

J. Griesemer (✉)
University of California, Davis, CA, USA
e-mail: jrgriesemer@ucdavis.edu

S. Leonelli et al. (eds.), *Life Beyond Human Boundaries: John Dupré's Philosophy and the Interdisciplinary Study of the Life Sciences*,
https://doi.org/10.1007/978-3-032-26977-5_10

work is to try to say what some of his ideas mean to me and for my perspective(s) on science and philosophy of science.

We came to somewhat comparable views on metaphysics and epistemology of (and for) science more or less in parallel, no doubt aided by conversations we have had over the years, beginning with his time at Stanford when I was up the road at Davis. We came to our views through rather different pathways, with different philosophical and scientific exposures, contexts and strategies. These background differences became clearer after he moved geographically farther away in his return to England. Curiously, we interacted more frequently when we were situated across an ocean than when he was close by at Stanford.[1] Those contextual differences and changes explain some of the differences of philosophical perspective I want to articulate.

Among his generation of philosophers, John has been a leader in taking up and making up new topics for philosophical investigations of science: at the intersections of science & values, in promoting microbe-centered[2] philosophy of biology, and especially in developing themes of disunity with other "Stanford school" philosophers challenging received views of the sciences as aiming to produce unified knowledge of the "natural" world. John has been especially tenacious in tracing the disunifying consequences and philosophical impacts of "classificatory pluralism" on scientific practices, philosophy of science and philosophy generally—themes he has championed through his important project on the disorder of things and the disunity of science (Dupré, 1993, 2012). My goal in this essay is to try to push him a bit on consequences of classificatory pluralism for the tenor of his metaphysical stance on process ontology.

As John's work on these topics developed and I pursued others, I was content to be a follower. I have always thought that our broader and deeper commitments and topical interests ran somehow in parallel, especially on process ontology, though he was articulate and head-on about it and my views were more in dialogue with the developmental systems theories of Susan Oyama, Paul Griffiths and Russell Gray (e.g. Oyama et al., 2001). All along I imagined that I would simply "circle back" one day to consider each of his specific arguments in the detail they deserve. His prolific writing has made that a very hard goal to meet and one I am no longer sure is necessary: "circling back" has become more of a homeorhetic rodeo, jumping from one fast-moving horseback to another, seeking resonances, than a simple linear read-and-respond as John's work cantered along. There is so much scholarship now on each of the topics John has been instrumental in opening up that there is just too much to mention, let alone think about addressing, in a single essay. So, I will focus on what I think have emerged for me as a few key differences or branch points. These are points that may have some bearing on the shape of processualist research programs

[1] John always extended an open invitation to visit him and Egenis at Exeter, though I must also thank Sabina Leonelli and Chee Wong for actually working out several happy invitations and visits.

[2] I learned not to be human-, mammal- or vertebrate-centric in my philosophy of biology initially from David Hull, but to take a positive turn toward microbes from John and his colleague Maureen O'Malley.

and might be fruitful (for me at least) to pursue further. I hope this kind of attention will honor John's work and perhaps serve to argue that it is of enduring value, even if John himself decides to spend more time tending his garden than writing philosophy about science.

10.2 Background Knowledge

One source of difference between us is that John actually trained in philosophy while I trained in "undeclared major" and then in genetics, followed by graduate school in an undisciplined inter-discipline that is in some ways adjacent to, yet in some respects distant from, philosophical tradition: "Conceptual Foundations of Science."[3] I have also been more sympathetic to the abstractions of the population-level sciences than John has, perhaps due to my upbringing in the Chicago school biology of Sewall Wright (population genetics), Thomas Park (population ecology), Ross Kiester (population demography, behavioral ecology and evolution), and Michael Wade (population-level evolutionary biology and group-level selection). That has given me perhaps a certain degree of interest, if not sympathy, for abstract population models, even in economics, though like John I have little sympathy for what policy-makers and talking-head policy-fakers sometimes do with economic and population-genetic models.

I'm not sure this is a substantive difference between us, but it seems that John's philosophical strategy is more addressed to taking models at face value for what they assume and say, such that he focuses his attention and pressure on reasons and choices in making and considering such models in the first place. If evolution does not optimize, why make or take seriously optimality models? If some organisms are diploid, why make or take seriously haploid genetic transmission and population models for them? If natural systems are mostly far-from-equilibrium, why make or take seriously equilibrium models for systems like that? If people classify differently, for different purposes, why think people should or would sort kinds in the same ways?

These are good questions, but John seems more focused than I am on the implications of such models in both science and philosophy for reasons not to pursue them and to adopt alternatives, while I have been more focused on practices, methods and means of building or using such models along with assessing appropriate conditions, standards, and uses when applying them. My training with Bill Wimsatt may have made me more open to models that at face value are blatantly false, in the interest of seeing what and how scientists and policy-makers do things with them, whether for good or ill. Perhaps that has made me more of a bystander in the science & values movement, while John has sensed the urgency of its call to evaluate the pursuit-worthiness of certain kinds of scientific practice. I am in sympathy with John and

---

[3] The inside joke in CFS at the University of Chicago in my day was that "science didn't have any conceptual foundations." In other words, my PhD is in a subject that studies something that doesn't exist.

others who think that some kinds of science are really just not very good or indeed are pretty bad science. Policies based on such "rubbish science" are also immediately suspect, including much that passes under topical headings like "sociobiology," "evolutionary psychology," "race science," "creation science," "behavior genetics," "human population genomics," "genome-wide association studies (GWAS)," and "ancestry studies." But the kinds of models used in these practices are also used elsewhere to good or better effect, so study of the models and how, why and where they are used can be of philosophical value even when some uses are repugnant.

I agree with criticizing and rejecting projects like sociobiology or scientific practices of making up people based on thin characterizations of genomic ancestry or biological races, sexes or genders. I am concerned less with calling out pseudoscience (which I agree is necessary and important), than with understanding how false models might lead to truer theories. Sometimes it is important for policy reasons to reject science built with false models, but studying false models as means to bad or worse policies is not the same project as studying false models as means to truer theories (Wimsatt, 1987).

Our differences in training backgrounds may also explain why I have not felt the force of disagreements as compelling choices among philosophical alternatives, as they sometimes appear to me in the way John frames some of his arguments regarding major philosophical dichotomies: unity versus disunity of science, monism versus pluralism, substance-ism versus processualism. Maybe that's my reading more than John's view of things. My strategy, following Wimsatt, has been to look for robustness across alternatives rather than seek the single one that best (or better) serves multiple contexts. I suspect John agrees that there is no single best or better philosophical view that serves all, or even many contexts. We are pluralists of a feather on that. I think our differences in background sensibilities stem mostly from differences of backgrounds anchored in philosophy or in science, which I think explains different inflections we give to our forms of pluralism as well.

10.3 (Some) Varieties of Pluralism

John and I are both "processualists," favoring a process metaphysics over a substance metaphysics for understanding the biological sciences. However, John presents this in terms of a choice between better or worse ontologies as ways to understand organisms (at least as I read Dupré, 2018 and Nicholson and Dupré, 2018). To a philosopher faced with an argument that one ontology is better than another, it would seem irrational to choose or commit to a worse ontology. Scientists wear their ontologies on their sleeves, and don't worry so much about foundational unities of ontology as about what moves the work forward. My hybrid sort of favoring is based on yet a different perspective, based on comparative study across ontological perspectives and pragmatic commitment.

Favoring one ontology over another can be framed in arguments that one is better than another in all contexts, as traditional metaphysics or first philosophy might suggest, or that one is better than another in some contexts, as John's arguments for favoring process ontology for biology may suggest, or that because context includes personal research choices (or better: commitments) to invest time and effort and background knowledge of researchers, a different strategy for seeking "generality" or "pattern" across contexts is needed.

Pursuit of, say, a processualist rather than substance-ist understanding of biological modeling, theorizing, practices, and construction of empirical phenomena out of human engagement with/in "nature" can make sense for one researcher and not another. Context includes variety of personal context of training, skills, resources, and interests. Research is also typically conducted in the context of a community. I favor an ontology that is different from that of others in my community so as to have multiple ontologies "on the table" for consideration and comparative analysis. Probably if most philosophers of biology were convinced by John to become processualists, I would re-commit to substance ontology, just to help keep it alive and on the table for consideration. This communitarian attitude is the kind of pluralism promoted by John Beatty (e.g. 1997) for understanding arguments in biology. Here, I extend it to philosophy of biology as well, but with the twist that there is need of diversity of metaphysics to pursue generalization in philosophies of sciences.

I think there is a difference of emphasis in our expressions of pluralism for philosophical practice. I chose to adopt a process ontology because there are plenty of substance-committed philosophers of science already. My goals are somewhat differently oriented than John's, so our strategies are also different. He argues that process ontology better illuminates biology than substance ontology. I seek understanding of biological practices that are robust to alternative ontological stances of philosophers, not the best ontology for biology. My view is "multi-perspectivist." My commitment need not be everyone's commitment. Indeed, I think it would actually be bad for philosophical understanding of science and for science itself if everyone committed to the same metaphysics. While I think process metaphysics is a good turn to take, I hope the result is neither than everyone jumps on board, nor that we just end up with another unresolvable dualism: process versus substance.

A multi-perspectival stance can be taken as communitarian, e.g. each researcher may accept or reject John's arguments favoring process ontology for biology, but the community of philosophers of biology might in addition pursue robustness analysis comparatively across ontological perspectives due to the varying commitments of a variety of philosophers. On the other hand, a multi-perspectival stance can also be something more personal.

In my view, a philosophical account of some aspect of biological science is not adequate if there isn't at least a possibility of robustness analysis for that account, in that philosophical project, in the local context of that investigation. Indeed, I take multi-perspectivism to be valuable as both a personal and as a community stance. At the personal level, a focus on process, say, can usually be tempered and balanced by considering alternative ontological perspectives in that project, just as scientists who weigh alternative models and modeling strategies for a given phenomenon can consider trade-offs or differences among model virtues, as argued by Levins (1968) for ecological models (cf. Griesemer, 2018). At the personal level, robustness is the goal, so scientists who consider multiple models have a shot at assessing it within projects. Similarly, philosophers who consider multiple ontological perspectives within projects have a shot at assessing robustness of their philosophical findings. What makes this personal level stance interesting is different than simply the community saving us all from personal-level bias. Rather, it reshapes how to think differently about goals and methods, models and perspectives, as well as what is pursuit-worthy in light of pragmatic interests and social values than can commitment to a single best (even if only best for me) perspective.

Generation of a diversity of perspectives in search of robustness to idealizing assumptions of ontologies, together with idealizing assumptions of models within perspectives, is a strategy distinct from generating diversity for the sake of choice of best or better fitting model and best or more apposite perspective. One way to be a multi-perspectivist is to support community pluralism among projects and contexts. Personal pluralism in addition is more demanding. It requires considering multiple ontologies within a single project in a single context, which means individual researchers must embrace, i.e. entertain and compare, multiple ontologies.

John's pluralist stance seems more like a variety of contextualism about metaphysics at a community level—a monism of the moment—than multi-perspectivist, though allowing for community-level ontological pluralism.[4] Although he suggests that substance or thing ontologies may have a place, occasionally, and for some kinds of problems, in his recent work he seems in favor of a process ontology on the whole, for biology at least, as generally preferable, so perhaps John is even showing tendencies toward metaphysical monism for philosophers (process, not substance, for all) even while endorsing pluralism for scientists and about science. In the context of another science or a different set of research problems to those John considers in biology, or perhaps in tolerance of unpersuaded philosophers of biology, maybe a substance ontology would be better suited to them, hence he might endorse a philosophical pluralism for all sciences collectively.

Multi-perspectivism applies to philosophy as well as to science and, applied at personal level, to me as well as to you. This reduces the status of metaphysics from "first philosophy" to meta-theoretical perspectives on philosophical strategies that are pragmatically worth considering (whether by an individual, group or whole community of researchers). A contrast between John (or at least, the way I have been reading him) and me, is that I seek robustness across metaphysical perspectives in my own projects while John seeks to evaluate choice among them, if not for him or me, then for the community.

To my mind, multi-perspectivism is just a different way of articulating a joint disunity of nature and disunity of science thesis. Rather than unified philosophy

---

[4] I thank Natasha Haddal for helping me understand contextualism and pluralism as distinct projects: Haddal, N. "Bridging the Towers of Babel: Pluralistic Approaches to Understanding Biological Sex." PhD Dissertation, 2025, University of California, Davis.

for disunified science (and nature), multi-perspectivism seeks robustness and reliability across the range of particular metaphysical commitments of particular philosophers and local philosophical communities and their particular projects in particular contexts of inquiry for particular purposes. Just like science.

Put differently, philosophers and scientists are in similar positions. Practitioners in each must make metaphysical choices—commit to an ontology—for the pragmatic sake of practicing inquiry within a diverse thought community. Whether inquiry leads to competitive solutions among which a choice must be made or cooperative solutions in comparisons that collectively offer different respects and degrees of robustness depends on the stances people take within their communities. The plurality of classifications of scientific kinds in sciences like evolutionary biology, as discussed further in the next section, applies to metaphysical kinds as well for philosophical communities; to many philosophers, both substances and processes seem important to various philosophical projects.

John's work traces consequences of classificatory pluralism for scientific inquiry. In shifting attention from the diverse models scientists make and use to understand the natural world to the diverse metaphysical perspectives philosophers of science make and use to understand the worlds of scientific practice, I am pushing for a disunity thesis regarding philosophy of science in practice, which in turn should raise at least an eyebrow about the way John's arguments favor process ontology for understanding the biological sciences.

There can be multi-perspectivist approaches in philosophy of science as well as in science that are distinct from pluralisms about ontologies for science that are monistic for philosophers of science. In my view, there is likely no best, all purpose metaphysics any more than there is a best, all purpose model, even for a single context of inquiry (Levins, 1968, p. 7). And that goes for inquiry within each philosophical project, not just among them. It is incumbent on philosophical as well as scientific researchers to assess the robustness of their findings to their own idealizing assumptions, not just to those of others. Classificatory pluralism (here about ontologies) extends disunity arguments to philosophy of science and philosophy generally.

A community of philosophers of science ought to at least try to understand science in ways robust enough to stand up to the particular metaphysical commitments of particular philosophical practitioners. The same goes for scientists and their projects. Robustness of understanding, which I take to be a kind of robustness across philosophical and scientific-theoretical perspectives, parallels Bill Wimsatt's philosophical articulation of robustness across kinds of empirical models, a la Richard Levins (see Wimsatt, 2007). A multi-perspectivist should embrace, i.e. actually engage with, compare, and pursue robustness analysis over a collection of perspectives, rather than merely tolerate others for the sake of determining which is better or best. A processualist who is also a multi-perspectivist recognizes the zeal that comes with monism. A substance-ist who is also a monist is likely to push the heuristic assumption that substance ontology is right or better for biology further than a process ontologist would push it in order to favor a process ontology, and it is only by pushing hard does one discover failure points that can put both people and their communities onto a path toward truer theories. It remains to be seen whether understanding that is robust to perspectives leads to better understanding. My bet is that it will. As a multi-perspectivist, though, I also have to embrace the views of those who think it won't.

10.4 Statisticalism and Organicism

John and I both favor "organicism" in contrast and complement to mechanism in framing a suite of questions around what makes living "systems" distinctive empirically and conceptually. Organicist ideas help understand why theory- or explanatory- or methodological- reductionisms as ways of interpreting the biological sciences in terms of physics and chemistry, or the morphological in terms of the molecular within biology, really are not very helpful ideas for understanding organisms.

Organicism points to what is lost on reductionist pictures of organism. John's astute observations and clearly articulated arguments against reductionism in philosophy of biology hinge on recognizing that there are two "senses" of organism (Dupré, 2018) in play across the biological sciences: (1) a lineage concept of organism as part of a lineage of cells (formed from an initial cell such as a zygote), and (2) a living individual that behaves, develops, and engages in ecological interactions.

These two senses mirror distinctions biologists have made between so-called "ultimate" and "proximate" biology and their causes (Mayr, 1961). The two senses of "organism" drive wedges among classifications of organisms into kinds, which led John to the "classificatory pluralism" he developed in his disunity project (Dupré, 1993). In that project, it emerges that there can be no single, best, all purpose classification system for organisms. This pluralism in turn supported John's turn to processualism, through recognition that a substance ontology is ill-suited to a world in which classificatory pluralism holds sway.

John's classificatory pluralism offers a path to processualism quite different than mine, though I think both paths lead to a picture of epistemology entangled with ontology. My path developed from beginnings based on metaphysical-ish pluralism about kinds of causes (and kinds of causal theories), grounded in Cartwright's treatment of Aristotle's four causes (Cartwright & Mendell, 1984, Cartwright, 1989, cf. Griesemer, 2005). "Metaphysical-ish" because mention or appeal to the four kinds is required for ideally complete causal explanations, where explanation is a human, "epistemic-ish" project. I viewed Cartwright and Mendell's metaphysics-of-science application of Aristotle as a template for other theoretical ontologies or "meta-theories" framed within particular sciences. Weismann's nineteenth century theory of heredity, for example, manifests a "plasm ontology"[5] supporting germ/soma, gene/character, and genotype/phenotype distinctions in turn-of-the twentieth

[5] It might be interesting to explore the history of plasm ontologies at the turn of the twentieth century as a means of exploring the dance of substance and process in many kinds of emergent scientific disciplines including modern chemistry, physics and the social sciences as well as the biological sciences.

century biology (see also Griesemer & Wimsatt, 1989, reprinted in Ankeny et al., 2025).

John's path seems to have led him to greater skepticism than I hold toward the value of population-level sciences (biological "plasma" flows at every level of biological organization). To John, population sciences seem to seek to reduce macro- (population) phenomena to micro- (molecular) phenomena. To me, population sciences just seem to reframe questions about ontologies that (more?) easily count things at lower levels into questions about how to individuate flows at higher levels and vice versa.

Reduction from population to gene in evolutionary biology, as in Dawkins' well-known arguments for a "gene's eye view" and "extended phenotype" (Dawkins, 1976, 1982), are level-skipping explanations. In between the levels of populations and genes lies the level of organisms. Genes (or rather nucleic acid sequences) are compositional parts of organisms and organisms are compositional parts or members of populations (and species and ecosystems). That compositionality allows Dawkins to abstract "gene pools" as populations of genes that skip over the level of organisms in explaining how, why, where, and when genes have "replicator power," even if he must assume organisms that can and must mediate the compositional relations between gene and population levels. But where John sees reductionism as the main pernicious strategy of level-skipping explanations like Dawkins', I see other projects aimed at multi-level level-connecting explanations framed in terms of the same level-skipping accounts, such as Sewall Wright's insertion of population-level structure (Wright, 1931) into the macro-to-micro statisticalism of Ronald Fisher's "Genetical Theory of Evolution" (Fisher, 1930).

The statistical modeling project of representing population-level evolution as change in gene frequency[6] relies on tallying genes within and among organisms in John's first sense: genes are the "germ-line" lineage-forming entities or processes while, on Dawkins' account, somatic organisms are not lineage-forming (except by proxy). Rather, somatic organisms are outcomes of the "active" powers of genes. The level skipping is due to abstraction from the compositional features required for "active" powers to be linked with "germ-line" powers such that genes can play both roles. In Dawkins' replicator abstraction (genes are active, germ-line replicators), organisms are simple statistical aggregates that aid in tracing lineage relationships of genes (through Mendelian calculus or cellular mechanics) in order to track gene frequency change in populations. Statistical aggregation as the means of discovering population properties robs populational "entities" of any (or much) chance at causal powers of their own: active powers only extend power from genes to organisms.

---

[6] Dobzhansky's (1937) "reluctant sign of equality" between micro and macro evolution—gene frequency change and species evolution—became dogma in the modern evolutionary synthesis. Another way to express what I see as a difference between my approach and John's to population genetics level-skipping is that I focus on Dobzhansky's reluctance while John focuses on the dogma.

Dawkins' lumbering robots lumber[7] because they are products of the development of morphology, physiology and behavior into which genes extend their "active" powers via phenotypes and environments (extended phenotypes). They are organisms in John's second sense, where the relevance of genes riding inside is how genes contribute to development rather than to evolution. The reductionist explanatory project is to link population to gene by considering organisms to be merely the developed "lumbering robots" for the genes riding inside them, no more than the bags in what Mayr once called "bean bag genetics." On the other hand, all the development, physiology, morphology, and behavior of the robots themselves is merely the proximate means of ultimate replicator germ-line power. The proximate means of replicator active powers to engage environments merely point to the "forces" that feed back statistically to genes as selection differentials assigned to genes. Thus, on the Dawkins' level-skipping reduction, a population is nothing but a statistical aggregate of genes: a gene pool. For all that matters to the statistics, genes swim naked in gene pools. Their organisms are merely bathing costumes. Organicism seeks to study bathers rather than bathing costume fashions.

The plurality of organism concepts revealed in John's way to organicist processualism is through his classificatory pluralism. Another way is through my kind-of-causes/kind-of-theories pluralism. Both can be waved away by monist-reductionists (genes are the only[8] entities with real replicator powers) as a smoke and mirrors complication.[9] Only efficient, formal causes of gene frequency change matter to gene statistics. Active powers are hidden away in tangential material and final causes. Only a lineage concept of organism matters to gene statistics. Active powers are hidden away in "forces." In any case, Dawkins thinks what goes on at the organism level was not a genuine problem in the classical era of debates over "the units of selection" in the 1970s-1990s, just a reflection of "Necker-cube" pluralism of two ways of counting genes, where it is a matter of inconsequential preference or convenience or taste how one views the organism.

The snag is that gene to organism relations seem to require one sense of 'organism' while organism to population relations require the other sense, neglecting translation problems between the two senses of organism. Reduction cannot work[10] because there are no "bridge laws" translating the lineage concept of genes aggregated into organisms aggregated into populations into the behavior-morphology-physiology-development-ecology active powers concept of organism as "manifestors of adaptation" at multiple levels (see Lloyd, 2021, 2024).

---

[7] Lumbering is a metaphorical way of conveying that organisms don't have agency of their own (no "active powers") such that they might self-organize into groups that gain agency through social organization. So the lumbering of organisms breaks the link between John's two senses of organism. All powers rest with the replicators.

[8] Besides Dawkins' other replicator discovery: memes.

[9] I suspect as well that I was more concerned with group selection and multi-level selection than I think John was because the dismissal of "Wrightian" group selection by "Fisherians" was always a topic of concern in the kinds of causes arguments I interpreted to stem from "Weismannism" and the origins of substance/process dualism in "plasm ontologies."

[10] The critique goes all the way back to Hull (1974).

For John, classificatory pluralism is at the heart of reduction- (and monism-) defeating arguments. For a multi-perspectivist, the issue can be framed differently: given that 'organism' figures in both perspectives, how do we conceive of organisms in ways that reconcile (or at least facilitate comparison and robustness analysis among) ontologies, epistemologies and methodologies that seem to require a plurality of organism concepts?

Dawkins argued that he is a gene selectionist, not a reductionist, because in articulating his views he "rediscovered" the organism as concentrator of replicator (gene) power (Dawkins, 1982). But John's arguments pull back the curtain to show that the kind of ontology Dawkins needs to make that distinction must translate between the two senses of organism on pain of an incoherent or incomplete ontology. Dawkins' monism about units of selection in evolutionary biology evades the problem by waving away the organism as mere consequence. Despite Dawkins' protestations, his selectionism amounts to a reductionism because it is built on the statistical platform of a substance ontology of genes, where organisms and populations are (merely) aggregates of countable things. How active powers are exercised is a side issue for "proximate" biology to consider, not a core issue for "ultimate" evolutionary biology aimed at universality.

Contemporary organicism as a movement takes organism as a level in its own right seriously, shifting focus on genes as well: from genes as counting units in a mathematical calculus of statistical population studies to genes as things in their substantial "physiological" roles in development and metabolism. John recognizes the value of a statisticalist approach for a very few biological questions and favors an organicist approach for most. I am an organicist in the same way that I am a processualist: as a multi-perspectivist who seeks understanding of genes, organisms, and populations that is robust to a plurality of ontological (and indeed epistemological) commitments and perspectives. I think the difference is that John seems to take 'population' at face value from the models as (mere) statistical aggregates, while I accept the challenge of trying to understand how they flow. I suspect John's pervasive ontology would lead him to agree that if populations are anything (other than mental constructs), they must be processes. If so, the ontological challenge is to understand that, and then to reframe the problem of multi-level biology.

Multi-perspectival pluralism about these thing-processes seeks to understand what this plurality of senses of 'organism' can reveal about how biology is practiced, both in lineage-mode and in active-powers-mode at every scale and level. Practice must face up to the fundamental facts of variation within and among diverse multitudes, among the biological phenomena and among the philosophical perspectives taken to study them. Progress toward a multi-perspective philosophy of biology might be made by reimagining population- and gene-level modeling projects as a multi-scale problem that links (if not integrates) all three scales of molecule (and not just nucleic acids), organism, and group. That would be a practical, procedural integration as conceptual-theoretical linkage, rather than a theory reduction to lowest level substances or processes of the traditional sort. The looser senses of linkage, intersection, or "cross-cutting" across scales needed for such a project, compared to the bridge laws linking explanatory theories for/at different levels, connect well with conceptions of multi-scale modeling in the physical sciences (e.g. Batterman, 2001; Bursten, 2018; Winsberg, 2010).

10.5 Representationalism and Navigationism

A third potential difference between my views and John's regards a topic more "epistemological" than "metaphysical"[11]: the commitment to representationalism in philosophy of science. John does not address this topic directly (at least not in his work on disunity, reductionism, pluralism, and science & values) so far as I am aware. But it lurks, so I suspect he is a "representationalist"—someone committed to the presupposition that philosophy of science must/should be built on a platform for interpreting sciences as "representing" (phenomena in, aspects of) "nature." Representation can support realisms, relativisms and everything in between, so long as some sort of correspondence or similarity relation between mind-dependent entities (like hypotheses, models, theories) and nature-dependent entities (like noumena or phenomena) holds in consequence, such that something like "evaluations of fit" can be made.

Representationalism is at the root of many forms of epistemology/metaphysics dualism. The representations are "epistemic objects"—the concepts, models, hypotheses, theories, predictions, explanations, or visualizations scientists conjure and use in their work, while phenomena are manifestations of objects characterized in some particular kind(s) of way(s) so as to "reflect"[12] an ontology—a list of the kinds of "things" or "processes" or whatevers there are "in nature."

Reduction fails for biology because organisms must be classified in different ways, as different kinds of kinds, in the contexts of different problems and projects. The failures, in my view, turn out to be failures of representation in one way or another: failure to represent phenomena accurately, precisely, or resolvedly so as to distinguish what (things) should be distinguished, whether in models, definitions, empirical claims, hypotheses, or classifications. In turn, representation is key to understanding what can be meant by success in prediction, explanation, understanding, or control of "nature" or "its" empirical phenomena—anything that involves judgments regarding relations of fit between theory and nature. Representations can be judged to be accurate or distorted, precise or imprecise, highly resolved or vague. Monisms call for choosing

[11] I often say I never met a dualism I could trust. The dualism of epistemology and metaphysics, knowledge and ontology, is one I not only distrust, but find to be an obstacle to progress in philosophy of the sort Bill Wimsatt sought in his work on the ontology of perspectives in the context of his "engineering philosophy of science." See especially the essays in Sect. 10.3, "Reductionism(s) in Practice" in Wimsatt (2007).

[12] "Reflection" and other visual/optical metaphors for similarity or resemblance are abundant among representationalists. Interestingly, some science studies scholars turned to "refraction" to resist pure representationalism and capture a sense of the co-constructive entanglement of the "object" (mirror, water surface, diffraction grating) bouncing back to the "subject" (human investigator) a modified image.

the "best" (kind of) representation. Monisms of the moment call for choosing what's "best" (or good enough) in the moment, given a context of inquiry, for the job. Some pluralisms are monisms of the moment, assuming choice contexts at different moments are simply incommensurable, such that there just are different contexts at different moments in which different choices are (relatively) suitable. Comparison of different choices for different contexts in different moments is not itself a question worthy of consideration or investigation. Evaluations of fit happen within a context and once evaluation has been made, that's the end of that story.

I used to be a representationalist when I focused on models in biology, but I have moved away from assuming representation was what conceptualizing and modeling are always for. I am some kind of pluralist about that too. What are some alternatives? My sociologist colleague, Elihu Gerson, holds more of a characterize-and-compare (for analogies and contrasts) perspective than a represent-and-test (evaluate for fit) perspective. I'm sympathetic to that alternative as "better" at interpreting many biological sciences for some kinds of problems and projects in some kinds of contexts of inquiry. The perspective I aim for is different still, more "processual" and emphasizing navigation-and-narrative: navigating conceptual, theoretical, and empirical "landscapes" (e.g. Griesemer, 2017, 2018) and offering journey narratives of empirical travels—finding things out—in cases of scientific practice guided by conceptual "compasses" that work to guide rather than representations that fit.

Compasses are tools for orienting or guiding journeys through landscapes, without (in some cases) necessarily needing any kind of representation of the landscape. To navigate by compass, one only needs to commit to the notion that the landscape is subject to navigation-by-compass. That commitment may depend on a background theory (e.g., magnetic compass aligns with Earth's magnetic field), or it may not (e.g., the compass always points toward the pole star). Representation practices can be products of navigation and narrative, e.g. making maps or navigational charts upon which to trace journeys and navigate them at "waypoints" (Winther, 2020).[13] In turn, maps may be used as navigational aids in their own right, sometimes complementing, sometimes supplanting compasses.

Models can be used as compasses, but the typical reading of 'model' by philosophers of science is as representations (e.g. Weisberg, 2013). Compasses, like models and other heuristics, can fail or be made to fail for systematic reasons. Studying the systematicities of failure is a means to building "truer" theories (sensu Wimsatt, 2007, ch. 6). So, now we have three perspectives to consider at this "epistemological" (sic.) "level": representationalism, comparativism, and navigationism.[14]

[13] Winther (2020) offers extensive discussions of map making and how map abstraction is linked to ontology.

[14] Again, as a multi-perspectivist, embracing one perspective is not a rejection of another. I seek robustness across perspectives. Just as I think there are more than two ontologies, substance and process, to consider for ontologies, I think there are more than two epistemologies to consider. I resist choosing among among substance and process ontologies by considering structure-substance ontologies, capacity-function ontologies, and activity-process ontologies. I resist choice among competing epistemologies by considering more than two kinds. I take it that representationalism, comparativism, and navigationism are philosophical perspectives on the project(s) of epistemology.

If John were a representationalist, I would expect him to either defend his choice of presupposition or develop an argument favoring another epistemic perspective. This is the basis on which I maybe slightly rudely accused him of being a monist at the book celebration event for his superb volume co-edited with Dan Nicholson, Everything Flows: Towards a Processual Philosophy of Biology (Nicholson and Dupré, 2018). They recognize (see also Dupré, 2018) processualism as a "better" ontology for biology, which makes them monists of that moment, in which some problems they identified may be better understood as problems with a process ontology than with a substance ontology.

I would not defend my personal philosophical strategy and perspective choices to be guided by an alternative presupposition about epistemology any more than I would about ontology. Navigationism is not "better" than "representationalism" because it is unclear how to apply judgments of better and worse to alternative presuppositions to representationalism at this "level" (i.e. epistemology rather than (sic.) ontology"). Nor is navigationism simply different than representationalism and better in some contexts, worse in others. The point is not epistemological monism of the moment. For a multi-perspectivist seeking robustness among perspectives, judgments of better or worse "fit" of epistemologies is not an option because evaluations and judgements of fit are epistemic activities. We can't use representationalism to ask for reasons not to be representationalists. The kind of understanding I want to explore through a navigation-and-narration epistemology is what one comes to understand (about oneself as much as about what is seen, witnessed, experienced) from a journey. There need not be any "representation" involved.

A different angle on John's epistemic perspective appeared in his collaboration with Sabina Leonelli toward a "process epistemology" (Dupré & Leonelli, 2022). They argue for an epistemology in the service of a process ontology. They focus on reification practices.[15] "Reification" consists in the identification of more or less stable features of the flux, and treating these as constituting stable things." This addresses my kind of pluralism, multi-perspectivism, but from a fully epistemic perspective, by describing a process by which which stable "things" are reified out of processes, which is to say that "things" are (merely) "treated-as", i.e. epistemic things, while the dynamic processes out of which things are reified, really are, i.e. ontic. That's a nice complement to John's process ontology, but it doesn't solve problems for multi-perspectivism about "ontologies" because to seek robustness among ontologies, you have to embrace and engage, not merely tolerate, more than one of them.

My strategy on epistemology is not to double the pluralism, for epistemology as well as for ontology. Rather, it is is to entangle them into onto-epistemology

When there are three items on the table, the project of searching for robust understanding starts to become interesting, because 3 items implies 3 binary contrasts and one 3-way comparison, so there are more comparative frames than there are generating items. Just as I find it interesting to pursue integrative structure–function, function-process, and process-structure ontological robustness analyses, I would like to study combinations of navigate-and-narrate/characterize-and-compare epistemologies as well as representational combinations of each of the other two.

[15] See also Winther (2020).

(Griesemer, 2021). For me, this involves a strategic shift to a Wimsattian philosophical perspective that blurs the epistemic/ontic dualism itself by treating perspectives as (partly) ontic and (partly) epistemic, whatever "partly" can mean here. Wimsatt's notion of a onto-epistemological "thicket" seems apt, if somewhat inscrutable (see Wimsatt, 2007, ch. 10, see also Griesemer, 2021). What I am trying to say here is that Wimsatt invented a new kind of problem of reconciling alternative ontologies by troubling the epistemic/ontic dichotomy: joining what we (can) know and what there is into one joint project rather than two separable kinds of projects. This is in contrast to assigning thing-classification to an epistemic category (reified process) rather than to a process ontology category (process) and a process epistemology category (reification), while holding metaphysical theorizing about ontology at bay from epistemology and conversely. The only glimmer of a juncture between them in this additive strategy is the common recognition that knowing-subject humans are part of the world of known-object processes, even while holding off the notion that human reifications are still somchow "ontic" (because knowing is doing and humans are processes too) in deeming them "epistemic" and subjects of "epistemology." At least, that's how I read their paper on process epistemology with John's process ontology as context.

One answer I might offer on why I am becoming a navigationist (navigator rather than modeler?) is that the philosophical journey I am on doesn't (yet) address research questions that can be centrally framed in terms of judgments of fit of concepts and models or perspectives and ontologies to phenomena. Judgments of fit are the core of evaluative concepts such as testing, confirmation, falsification, corroboration and the like that all operate on the platform of representationalism. My concerns are more exploratory and my accounts more tentative than all that, in journeys through conceptual terrain I am less confident of navigating smoothly. I'm not sure if this is simply a different frame for contextualist pluralism of the sort Bruno Latour recognized when he called scientists "Janus-faced": relativists in the heat of investigation, realists about settled knowledge (Latour, 1987). Here it seems I'm proposing navigationism for science in discovery and representationism for science in justification. Popper all over again? No! A Wimsattian strategy tracks "epistemic" and "ontic" activities together. The distinction between them is the grand idealization of philosophy into epistemology and metaphysics as different books on the book of knowledge.

The best I can do in engaging representationalism for now is to suggest representation-sketches as anticipatory. My particular journey is through landscapes of biological reproduction processes and phenomena, and even when it seems smooth and easy, as in the case we thought we all understood in some respects to some degree—human reproduction—I want to trouble it and raise doubts about how we think about it, what we saw when we whizzed by it looking out the window on journeys we thought were about representing it in models and theories. E.g., I aim to challenge orthodoxy on how to think about what counts as a reproducer (reproducing process): retroviruses too? In the same project that examines choices of default and alternative narrative "grain" when characterizing the "life cycle" of retroviruses like HIV (Griesemer, 2014). Narrative is a key aspect of the navigationalist perspective because alternative, even conflicting, narratives of a series of events are to be expected in story-telling by diverse story tellers, in contrast to the hoped-for singularity of the confirmed hypothesis or best-fitting model for a given empirical phenomenon.

Ontological commitment is like travel insurance: you get a smooth journey or you get your money back. I'm not promised either in the landscapes of biology and biological practice I aim to traverse, nor with the fiddly compasses I have built so far. Multi-perspectivism about onto-epistemology seeks that kind of assurance: robustness at a different level, that of perspectives rather than models, to ensure a smooth journey and a satisfying narrative, no matter how one navigates through a particular landscape of phenomena. The best I can hope for is a conceptual compass that somehow gets me through: a device (concept, model, map, strategy, practice, procedure) that gets me through even if it doesn't tell me exactly what I am seeing or what it is "like" while I travel.

10.6 Conclusion

I have outlined three topics where John and I might find disagreement: (1) how to be a pluralist, i.e. whether to be a monist of the moment and choose an ontology that is better or best for the problem at hand in that moment's context, or to be a multi-perspectivist, i.e. to seek understanding in narratives robust to alternative complementing "ontologies" as aids to navigating complex phenomena; (2) how to deal with phenomena/substances/processes at meso-scales to address questions of level-skipping explanations such as attempted reductions of macro (population) phenomena to micro (gene/DNA) phenomena while ignoring the "mess in the middle" such as multiple senses of 'organism'; and (3) representationalism as a reflection of epistemology/metaphysics dualisms that I think John may still operate with, that I aim to challenge with a Wimsatt-style multi-perspectivism.

In the end, I think it is wonderful that John and I agree about so many things and yet seem to disagree about them at the same time. I wouldn't dream of arguing over whether his approach, or view, or perspective, is better or worse than mine. I don't want to take myself too seriously, and certainly not more seriously than I take John, because I aim for understanding that is robust across both our perspectives (and those of many others, especially those who have collaborated with John on his many fruitful projects). As a multi-perspectivist, I can't take myself too seriously or I won't be open to the perspectives of others in the right way to carry out my projects. As a multi-perspectivist, I leave it to others, including John, to figure out whether or how to judge better and worse across whatever perspectival difference distinguishes what John does from what I do. Certainly, John's work is of the highest quality, which draws me to it regardless of topic, but as a multi-perspectivist, what is of most importance to me is to ask: better or worse for what purpose, for whom, in what problem-context?

A mark of the kind of process called "a good friendship" is that it persists robustly, despite and because of difference and disagreement as well as similarity and agreement.

References

Ankeny, R. A., Dietrich, M. R., & Leonelli, S. (Eds.). (2025). *Scaffolding: Selected contributions of James R. Griesemer to History, Philosophy, and Biology*. SpringerNature.

Batterman, R. W. (2001). *The Devil in the Details: Asymptotic reasoning in explanation, reduction, and emergence*. Oxford University Press.

Beatty, J. (1997). Why do biologists argue like they do? *Philosophy of Science, 64*, S432–S443.

Bursten, J. R. (2018). Conceptual strategies and inter-theory relations: The case of nanoscale cracks. *Studies in History and Philosophy of Science Part B: Studies in History and Philosophy of Modern Physics, 62*, 158–165.

Cartwright, N., & Mendell, H. (1984). What makes physics' objects abstract? In J. Cushing, C. Delaney, & G. Gutting (Eds.), *Science and Reality: Recent Work in the Philosophy of Science* (pp. 134–152). University of Notre Dame Press.

Cartwright, N. (1989). *Nature's capacities and their measurement*. Clarendon Press; Oxford University Press.

Dawkins, R. (1976). *The Selfish Gene*. Oxford University Press.

Dawkins, R. (1982). *The Extended Phenotype: The Gene as the Unit of Selection*. Oxford University Press.

Dobzhansky, T. G. (1937). *Genetics and the origin of species*. Columbia University Press.

Dupré, J. (1993). *The disorder of things: Metaphysical foundations of the disunity of science*. Harvard University Press.

Dupré, J. (2012). *Processes of life: Essays in the philosophy of biology*. Oxford University Press.

Dupré, J. (2018). Processes, organisms, kinds, and the inevitability of pluralism. In O. Bueno, R.-L. Chen, & M. B. Fagan (Eds.), *Individuation, process, and scientific practices* (pp. 21–38). Oxford University Press.

Dupré, J., & Leonelli, S. (2022). Process epistemology in the COVID-19 era: Rethinking the research process to avoid dangerous forms of reification. *European Journal for Philosophy of Science, 12*, 20.

Fisher, R. A. (1930). *The genetical theory of natural selection*. Clarendon Press.

Griesemer, J., & Wimsatt, W. (1989). Picturing Weismannism: A case study of conceptual evolution. In M. Ruse (Ed.), *What the Philosophy of Biology Is, Essays for David Hull* (pp. 75–137). Kluwer Academic Publishers.

Griesemer, J. R. (2005). The informational gene and the substantial body: On the generalization of evolutionary theory by abstraction. In Jones, M. R. & Cartwright, N. (Eds.), *Idealization XII: Correcting the model. idealization and abstraction in the sciences* (pp. 59–115).

Griesemer, J. R. (2014). Reproduction and the scaffolded development of hybrids. In L. Caporael, J. Griesemer, & W. C. Wimsatt (Eds.), *Developing scaffolds in evolution, culture, and cognition* (pp. 23–55). MIT Press Cambridge.

Griesemer, J. R. (2017). Landscapes of developmental collectivity. In S. B. Gissis, E. Lamm, & A. Shavit (Eds.), *Landscapes of collectivity in the life sciences* (pp. 25–47). MIT Press.

Griesemer, J. R. (2018). Mapping theoretical and evidential landscapes in ecological science: Levins' virtue trade-off and the hierarchy-of-hypotheses approach. In J. M. Jeschke & T. Heger (Eds.), *Invasion biology: Hypotheses and evidence* (pp. 23–29). CABI.

Griesemer, J. R. (2021). Levels, perspectives, and thickets: Toward an ontology of complex scaffolded living systems. In D. S. Brooks, J. DiFrisco & W. C. Wimsatt (Eds.), *Levels of organization in the biological sciences* (pp. 89–109). MIT Press.

Hull, D. L. (1974). *Philosophy of biological science*. Prentice Hall.

Latour, B. (1987). *Science in action: How to follow scientists and engineers through society*. Harvard University Press.

Levins, R. (1968). *Evolution in changing environments; some theoretical explorations*. Princeton University Press.

Lloyd, E. (2021). *Adaptation*. Cambridge University Press.

Lloyd, E. A. (2024). *Units and Levels of Selection*. https://plato.stanford.edu/archives/sum2024/entries/selection-units/
Mayr, E. (1961). Cause and effect in biology. *Science*, 134, 1501–1506. https://doi.org/10.1126/science.134.3489.1501
Nicholson, D. J., & Dupré, J. (Eds.). (2018). *Everything flows: Towards a processual philosophy of biology*. Oxford University Press.
Oyama, S., Griffiths, P. E., & Gray, R. D. (2001). *Cycles of contingency : Developmental systems and evolution*. MIT Press.
Weisberg, M. (2013). *Simulation and similarity: Using models to understand the world.* Oxford University Press,
Wimsatt, W. (1987). False models as means to truer theories. In M. Nitecki & A. Hoffman (Eds.), *Neutral models in biology* (pp. 23–55). Oxford University Press.
Wimsatt, W. (2007). *Re-engineering philosophy for limited beings: Piecewise approximations to reality*. Harvard University Press.
Winsberg, E. (2010). *Science in the age of computer simulation.* University of Chicago Press.
Winther, R. G. (2020). *When maps become the world*. University of Chicago Press.
Wright, S. (1931). Evolution in Mendelian populations. *Genetics, 16*, 97–159.

Chapter 11
Philosophy of Biology in an Interdisciplinary Key

Alan C. Love

Abstract A distinctive intellectual legacy of John Dupré and the Centre for the Study of Life Sciences (Egenis) at the University of Exeter is the productive interdisciplinary collaborations of scholars across the divide of different sciences and the humanistic study thereof. Here I honor that legacy by describing two veins of my own research that follow in these footsteps. First, I highlight a large-scale interdisciplinary project that interrogates one of the most controversial aspects of life science: teleology. Second, I detail local activities at the University of Minnesota through the Minnesota Center for Philosophy of Science that cultivate interdisciplinary intellectual community. These examples of institutionalized collaborative efforts illustrate the distinctive value of jointly exploring the assumptions and evidence behind scientific models to advance our understanding of how the sciences work and what they imply.

11.1 Doing Philosophy Interdisciplinarily

Analytic philosophy in the second half of the twentieth century (and into the twenty-first) has been dominated by conceptual analysis. The aim of conceptual analysis is to uncover the basic criteria of key concepts, such as causation or knowledge, that underwrite categorizing member instances correctly. In the context of philosophy of biology, concepts like GENE or SPECIES are natural targets for conceptual analysis. This approach to doing philosophy involves the intense scrutinizing of a variety of candidate member instances in diverse settings (e.g., is *x* or *y* a member of species *z*, and under what conditions?). The process of scrutinizing is often referred to as the method of cases. Although success in this endeavor is thought to be a distinct achievement, which is accomplished largely apart from empirical input and undertaken by individual philosophers (dubbed "armchair philosophy"), many have noted that it

A. C. Love (✉)
Department of Philosophy & Minnesota Center for Philosophy of Science, University of Minnesota, Minneapolis, MN, USA
e-mail: aclove@umn.edu

S. Leonelli et al. (eds.), *Life Beyond Human Boundaries: John Dupré's Philosophy and the Interdisciplinary Study of the Life Sciences*,
https://doi.org/10.1007/978-3-032-26977-5_11

has not been very successful according to its own aims. It is unclear whether there has ever been a single successful analysis of any concept (Huemer, 2015; Schroeter, 2004). However, philosophers have continued to defend the value of this approach, sometimes by more explicit appeals to contemporary psychological investigations of how we use concepts (e.g., Strevens, 2019).

Conceptual analysis and the method of cases have played a role in several debates in philosophy of biology, perhaps most famously in discussions of the concept BIOLOGICAL FUNCTION (Bigelow & Pargetter, 1987; Neander, 1991; Wright, 1973). However, many philosophers of biology (and philosophers of science more generally) have become less enamored with this approach and opted for a variety of other complementary methods, including historical or sociological inquiry, empirical analyses of concept use in scientific discourse, scrutiny of biological examples or theories, and direct collaboration with scientists. John Dupre was at the forefront of this alternative orientation, visible in places like his discussion of biological classification and natural kinds: "My strategy will be to compare the requirements of this analysis [of natural kinds] with some actual biological facts and theories" (Dupré, 1981, p. 66). Later he pursued close collaborations with sociologists like Barry Barnes to undertake historical investigation and empirical analysis of contemporary genomics (Barnes & Dupré, 2008). This represented, in part, fruits from the origins of Egenis, the Centre for the Study of Life Sciences at the University of Exeter (https://www.exeter.ac.uk/research/centres/egenis/), which began in 2003 as the Economic and Social Research Council (ESRC) Centre for Genomics in Society under John's leadership. It is distinctively *not* dominated by conceptual analysis and the method of cases but rather is unabashedly interdisciplinary, bringing together scientists, philosophers, and social scientists using diverse methods to better understand different aspects of the life sciences—from genomics and model organisms to data-intensive research and microbiology—and elucidate their ethical, social, political, and philosophical implications.

In both celebration and recognition of John's distinctive intellectual legacy and the productive interdisciplinary collaborations of scholars who intersect at Egenis, I describe two veins of my own research endeavors that follow in these footsteps. First, I highlight a large-scale interdisciplinary grant project that organizes different approaches to interrogate one of the most controversial aspects of life science: teleology. This project began in 2021 and formally ended in 2024, though it has ongoing strands still active today. Second, I detail local activities at the University of Minnesota through the Minnesota Center for Philosophy of Science that cultivate interdisciplinary intellectual community. In a leadership role like John's at Egenis, I have been directing these activities for more than a decade and therefore have firsthand experience with a cognate mode of academic operation. I intentionally ignore specific collaborative endeavors where I have worked jointly with one or more scientists, in part because others can share similar stories. The focus here is formal administrative dimensions of grant project or research center leadership that reflect combinatorial strategies of historical consideration, empirical analyses of concept use in scientific discourse, examination of life science practices, and different forms of multidisciplinary research teams. These examples of institutionalized collaborative efforts illustrate the distinctive value of jointly exploring the assumptions and evidence behind scientific models to advance our understanding of how the sciences work and what they imply. Overall, I hope to demonstrate that I hit similar notes to John in my philosophical work on the life sciences, doing philosophy in an interdisciplinary key.

11.2 Organizing Interdisciplinary Research on Purpose

Across the phylogenetic tree that represents the evolution of birds, a behavior called the "broken wing display" has repeatedly evolved. Its exact expression differs between taxa, but the basic activity is easy to recognize; Aristotle commented on this avian behavior in *History of Animals* IX.9. Some birds lay prone with both wings outstretched, while others favor a wing and emit a distress call (Fig. 11.1). The behavior has traditionally been associated with shorebirds but is not limited to any specific group or even to ground-breeding taxa. A recent study has shown that the behavior is present in no fewer than 52 families scattered across the phylogenetic tree, suggesting that it has evolved many times independently (de Framond et al., 2022).

The broken-wing display has all the hallmarks of a purposive behavior. Biologists can ask questions of the behavior like "what is it for?" or "what is its function or value?" and offer concrete answers (to protect offspring, to distract predators) that take their justification from historical processes of natural selection operating on ancestral populations. In addition, biologists might also be interested in investigating individual birds as autonomous, goal-directed, and self-maintaining agents. This might spur different kinds of equally legitimate questions: What kind of thing must a system be to engage in behaviors of this sort? What attributes must a system have to evolve such a behavior in the first place? Is there agency involved in the expression of the behavior, and if so, how is it structured? Does the display require the representation of a goal, and how exactly can we empirically detect this?

Questions like these steer us toward a wider space of possibilities for conceptualizing biological teleology and its empirical expressions. However, teleology—the study of or explanations in terms of goals and purposes—is treacherous territory for most biologists. Even though teleological reasoning is as old as any activity recognizable as biology (recall Aristotle's explorations of why elephants have a trunk-nose; Gotthelf, 2012), it has been subject to diverse and contradictory interpretations from the outset. Developments in biological theorizing, such as Darwinian evolution and mechanistic explanation at lower levels of organization (Richards, 2009; Riskin, 2016; Zammito, 2018), appear to challenge the significance and status of teleology, all while the theme of purposiveness has been salient from antiquity to the present day (viz., the broken-wing display) (Ruse, 2018; Walsh, 2008).

A fear or disdain of even talking about teleology is sometimes present in biological discourse. The evolutionary paleontologist George Simpson spoke of "the debased

Fig. 11.1 A North American killdeer (Texas, USA) engaged in a broken-wing display. Killdeers are plovers in the Family Charadriidae (plovers, dotterels, and lapwings). Ejmc, CC BY-SA 3.0 (https://creativecommons.org/licenses/by-sa/3.0), via Wikimedia Commons (https://commons.wikimedia.org/wiki/File:Charadrius_vociferus_tx1.jpg)

currency of teleology": "The words "finalistic" and "teleological" have, however, had an unfortunate history in philosophy which makes them totally unsuitable for use in modern biology" (Simpson, 1958, p. 520). And yet biologists have also recognized a deep tension surrounding teleology. J.B.S. Haldane purportedly remarked that "Teleology is like a mistress to a biologist: he cannot live without her but he's unwilling to be seen with her in public," but a more modern version of this tension surrounding teleology comes from Harvard evolutionary biologist David Haig:

> I was repeatedly warned against teleological thinking … by lecturers who spoke of the heart as a pump *for* the circulation of blood and of RNAs as messengers *for* the translation of proteins. …Concepts of purpose and function shape the practice of biology—this is the natural way to think about living things—but the overt language of teleology is censored. (Haig, 2020, p. 16)

The tension also appears in historical interpretations of biologists, such as the battles fought over whether Darwin was or was not a teleologist (Ghiselin, 1994; Lennox, 1993).

Thus, although biologists often use descriptive language that imputes purposiveness to living systems (agency, function, goal-directed behavior) (Okasha, 2018), many have argued that these conceptions are at best heuristic, and at worst egregious errors. However, two developments in recent decades have reinvigorated this issue. First, philosophers have shown that teleological concepts can be naturalized in different ways and thereby do not harbor inherent disrespectability in a scientific epistemology (Babcock & McShea, 2021; Forber, 2020; McShea, 2012; Mossio & Bich, 2017; Perlman, 2004; Wimsatt, 1972). The distastefulness of teleology as somehow tacitly representing outmoded theoretical commitments is no longer a genuine concern even if there remains hesitation among biologists in using the terminology, seen especially in the use of the allied term "teleonomy" (Dresow & Love, 2023). Now the primary challenge is sifting through the different options to see what empirical possibilities emerge.

Second, there has been a growing recognition that biological phenomena in the vicinity of teleology demand new conceptual frameworks that can translate into rigorous theoretical models and discriminating empirical tests. Many researchers are concerned with organisms as inherently agential or autonomous—even capable of mistakes (Hill et al., 2022, 2025)—and trying to capture this capacity or its downstream consequences in both experimental and theoretical models, especially evolutionary ones (Laland et al., 2017; Moreno & Mossio, 2015; Moreno & Peretó 2026; Newman, 2023; Rosslenbroich, 2023; Walsh, 2015; Walsh & Rupik, 2023; Watson, 2024). This concern links up with older traditions in cybernetics that attempted to apply systems thinking to biology and capture goal-directed behavior (Bertalanffy, 1952; Rosenblueth et al., 1943). Others have argued that new empirical findings prompt us to revisit our conception of mutational directionality, recognizing new categories not captured by the standard dichotomy between random mutation and Lamarckian mutation (Bolotin et al., 2023; Livnat & Love, 2024; Livnat & Melamed, 2023; Melamed et al., 2022, 2025). Still others have encouraged rethinking the directionality of processes occurring during ontogeny, such as in terms of bioelectricity (McLaughlin & Levin, 2018) or as evident in synthetic bioengineering with developmental constituents (Kriegman et al., 2021) or agent-like materials (Davies & Levin, 2023). Some have sought to characterize ecosystem organization or function (Odenbaugh, 2019; Sfara & El-Hani, 2023), whereas others have modeled the entire planet Earth in terms of self-organized processes with homeostatic properties conducive to living constituents (Bourrat, 2023; Boyle et al., 2025). Not all these calls are equally justified, and some will no doubt fade as time passes. However, the spur to this diverse work arises from genuine empirical advances, especially where features like organismal agency or mutational directionality have been overlooked due to blind spots in existing theoretical frameworks. Thus, it is poignant question how the biological research community should respond to such a situation.

What if a research program could be assembled where "purposive" phenomena remain central to biological inquiry despite significant and substantial disagreements about their status in evolutionary theory and mechanistic modeling? How might these phenomena be investigated scientifically? Can we simultaneously be careful and rigorous about the use of descriptive language that imputes purposiveness to living systems—and not merely as a heuristic convenience? What if teams of researchers were organized to approach these questions from multiple conceptual, theoretical, and empirical angles simultaneously? Could a systematic interdisciplinary conversation happen that brings a wide range of resources for talking about purpose into the context of model building, experimentation, and theorizing in biological investigation? Could philosophers potentially play a role in articulating richer and more precise frameworks for discussions of functionality, goal-directedness, and agency in biology that make assumptions in scientific research more explicit and thereby facilitate ongoing inquiry?

We answered these questions affirmatively in the grant project "Agency, Directionality, and Function: Foundations for a Science of Purpose" (https://archive.biologicalpurpose.org/; Fig. 11.2). More than 100 researchers on 24 different subaward teams were organized to address this issue through a novel, interdisciplinary, large-scale program that combines philosophers, theoreticians, and experimentalists to: (1) Articulate more precise concepts related to function and purpose; (2) Develop innovative formal models of agency; (3) Operationalize notions of goal-directedness for accurate measurement; and (4) Trial and implement methods and platforms to detect and manipulate directionality in living systems (Love & Dresow, 2022). The basic approach was to shift away from metaphysical questions ("what is teleology?") and instead focus on investigative and modeling practices: "how do biologists study agency, directionality, and function?," "for what aims?," "do their experimental strategies and modeling practices serve these aims?" In the helpful turn of phrase offered by Asa Gray in describing the significance of Darwin's work: "usefulness and purpose come to the front again as working principles" (Gray, 1876, p. 294).

Examples of the benefits of adopting a modeling perspective on how teleology is conceptualized and explained in living systems include increased attention to the ways in which researchers demarcate a system of interest from its environment, decompose those systems into different kinds of parts, and choose timescales on which systems or their components manifest purposiveness of different types. The broken-wing display in birds helps to illustrate each of these. Although we might intuitively treat the parental bird as the system and the environment as the offspring plus nest, predator(s), and physical surroundings, different modeling decisions arise from treating the system as a combination of parental bird and offspring plus nest or the environment as only the physical surroundings. Some of these modeling decisions relate to part-whole decisions for the system's parts (wing, reproductive system, vocal morphology, genetics) or properties (biomechanics, incubation duration, vulnerability, body mass, stress hormone activity, vocalization type), as well for the environmental parts (nest, other parents, ground cover, predator types, terrain types) and properties (nest cover, nest location, aggression/mobbing, hours of daylight, predator locomotion, hunting strategy, and visual acuity). Timescale considerations are pertinent depending on what kinds of organization for parts and their interactions are being modeled, such as visual detection and reaction (seconds), individual behavioral displays (minutes), frequency of predator encounter (days/weeks), and differential advantages to variation in display behavior (years/intergenerational). The modeling perspective makes evaluative standards explicit (Why is the system-environment boundary drawn in a particular way? What counts as a good decomposition of the system into interacting parts? How is the timescale for tracking a system's behavior

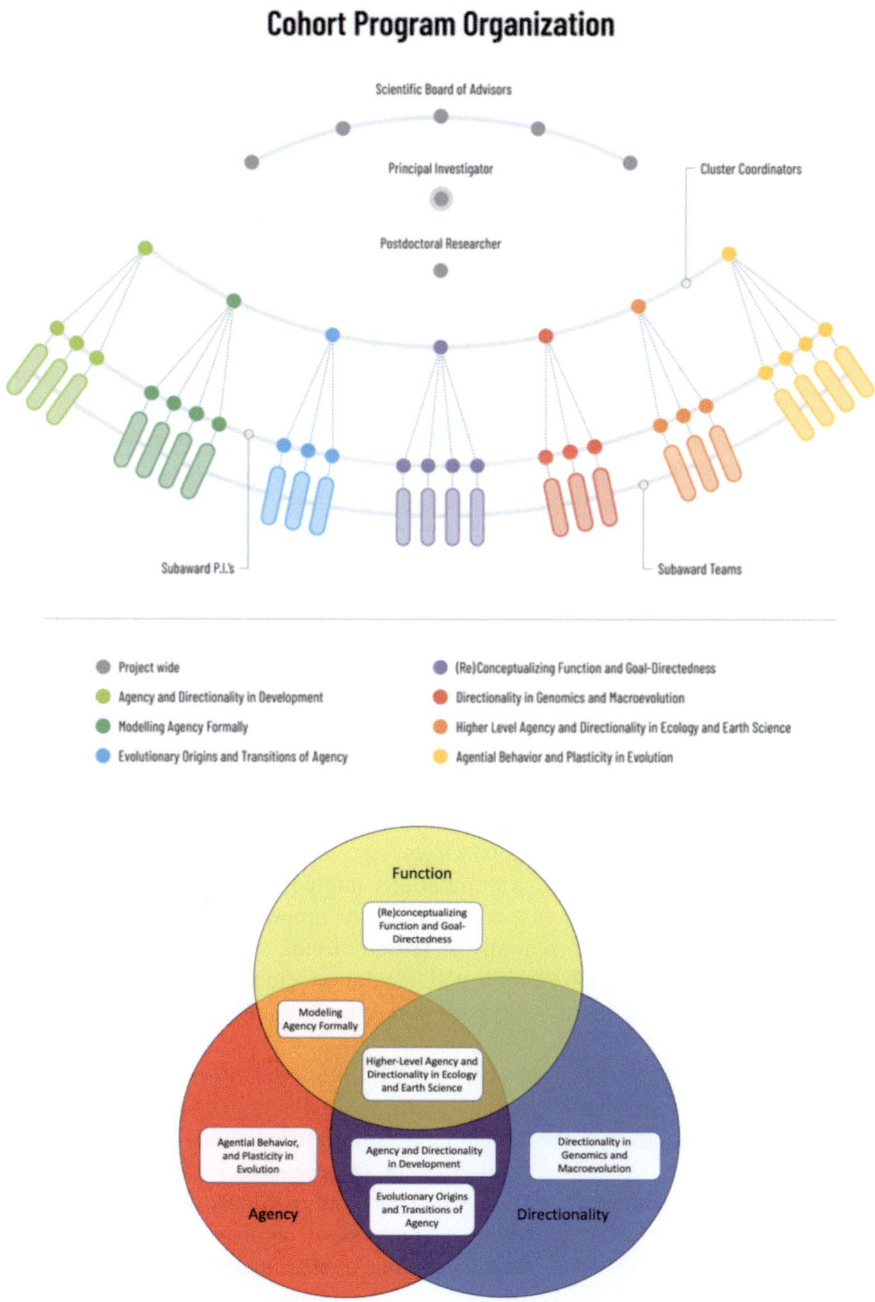

Fig. 11.2 Administrative organization of the Agency, Directionality, and Function grant (top). Thematic organization of the seven clusters that make up the cohort program (bottom)

selected?) and foregrounds both modeling decisions, such as abstractions and idealizations that generate trade-offs, and the diversity of teleological phenomena under scrutiny (e.g., behaviors exhibiting persistence and plasticity versus self-determining organization in contrast to basic functionality).

Over the period of the grant, a productive cycle of conceptual maneuvering emerged that can be summarized in the sequence: Embrace—Fragment—Characterize—Apply. First, shrug off "teleophobia" and *embrace* a broadly naturalized approach that jumps into the examination of purposive-seeming phenomena and the potential value of conceptualizing living systems teleologically. Second, *fragment* the concept of teleology into different interpretations or conceptions that correspond to different investigative aims and contexts. Many central concepts in biology (e.g., GENE, HOMOLOGY, TRAIT, and SPECIES) exhibit multiple non-arbitrary interpretations within and across disciplines, approaches, and research questions. Importantly, fragmentation does not automatically yield justification, as can be observed for other scientific concepts where particular meanings were jettisoned, and it is a conceptual maneuver with historical warrant (i.e., a reasoning strategy with a good track record).

Third, *characterize* these different conceptions, in part by elucidating distinct patterns of additional concept use ("conceptual neighborhoods"; Love, 2005) that correlate with the central concept of teleology. Relevant candidates have already been noted: AGENCY, CONSTRAINT, DIRECTIONALITY, FUNCTION, GOAL- DIRECTED BEHAVIOR, INFORMATION, MECHANISM, and SELF- DETERMINATION. These also can be subjected to fragmentation, encouraging a recursive application of these conceptual maneuvers (e.g., COLLECTIVE AGENCY, MINIMAL AGENCY, INTENTIONAL AGENCY; ACTIVITY FUNCTION, CAUSAL ROLE FUNCTION, ORGANIZATIONAL FUNCTION, SELECTED EFFECTS FUNCTION), which can be marked through the introduction of distinctions that reflect different commitments about system-environment boundaries, parts and wholes, and time scales (e.g., teleology is *intrinsic* if purposes arise from some set of properties or relations internal to a system; teleology is *extrinsic* if purposes manifest as a consequence of properties or relations external to a system) and applied differentially to specific living system types: organismal behavior (foraging, mating), metabolic self-regulation, embryogenesis, microevolution in populations, ecological succession and dynamics, and fossil record patterns (macroevolution).

Finally, *apply* or implement the characterization in modeling strategies and data gathering practices by operationalizing concepts and distinctions to uncover empirical signatures and elaborate theoretical models. How are boundaries drawn? What counts as parts and wholes and what kinds of organizational relations can obtain among them? What happens if you shift perspectives on spatial or temporal scales? This, in turn, facilitates the selection of abstractions, approximations, and idealizations that help isolate focal causal patterns, while recognizing tradeoffs that obscure some features to expose others. What sequences of modeling decisions and data gathering practices can be deployed in characterizing and explaining 'purposive' phenomena? Along the way, researchers more intentionally keep in view different investigative roles, methods or approaches, and research questions as these implementation decisions are made. As a result, evaluative standards related to these modeling decisions are foregrounded by differences in investigative aims (e.g., characterization, explanation, prediction, control), research questions (often guiding system circumscriptions and part-whole decompositions), and methodological orientation (e.g., formal or experimental).

The organization of research initiated within the "Agency, Directionality, and Function" project was intentionally a beginning, "Foundations for a Science of Purpose." The institutional scaffolding is currently being augmented and stabilized in a new consortium that will serve as a fulcrum to new research projects emerging over the next five years and beyond (https://www.biologicalpurpose.org/). The large-scale interdisciplinary architecture of the program was unique, and both its successes and failures have been the subject of separate sociological study so that lessons can be learned from it going forward. One of the most exciting features of the project was the cross-pollination of disciplines and teams that inaugurated a new community of research. At the core of this new community was the hard-won conviction that what we call "biological teleology" comes in a variety of manifestations and is associated with a plurality of constitutive and causal regimes. Consequently, multiple lines of inquiry that deploy different descriptive and explanatory resources and approaches are necessary.

11.3 Intellectual Community and the Minnesota Center for Philosophy of Science

Like Egenis at Exeter, the Minnesota Center for Philosophy of Science (MCPS) at the University of Minnesota is a research center that strives to explore the nature of the sciences (broadly construed, inclusive of mathematics) and any ethical, social, political and philosophical implications they might have within a setting that is self-consciously interdisciplinary. Its core aim is crystallized in the following slogan: "By interrogating the assumptions and evidence behind scientific models and characterizing the diversity of scientific practices, we advance our understanding of how the sciences work and decipher what kinds of answers they deliver" (https://cla.umn.edu/mcps). One of the central strategies for pursuing and accomplishing this goal is the cultivation of intellectual community across disciplines. This strategy goes back to the origins of MCPS in the mid-twentieth century and its connection to both Logical Empiricism and the Vienna Circle that came largely through its founder, Herbert Feigl (1902–1988).

In 1923, Herbert Feigl began his doctoral studies under the supervision of Moritz Schlick, a philosopher studying how early twentieth century developments in mathematics and physical science were transforming our understanding of what knowledge is (Edmonds, 2020). Schlick had secured the "Chair in the Philosophy of the Inductive Sciences" at the University of Vienna in the previous year and quickly became a nucleus of intellectual activity in a city bustling with cross-disciplinary conversations among mathematicians, philosophers, and scientists (Sigmund, 2017). Within twelve months, partly at the prompting of Feigl and other students, a regular discussion group led by Schlick, along with the mathematician Hans Hahn and the economist Otto Neurath, coalesced on Thursday evenings. It became known as the Wiener Kreis or Vienna Circle. Subsequently, the group was tightly associated with the philosophical approach labeled "Logical Positivism" or (as many affiliates later preferred) "Logical Empiricism." Feigl was largely responsible for solidifying this identity and appending the labels (Blumberg & Feigl, 1931; Feigl, 1943). As he remarked later in life, "the aim of this reflective enterprise was the clarification of the basic assumptions, concepts, and methods of the sciences" (Feigl, 1969, p. 633).

The Vienna Circle and Logical Empiricism had a major impact on intellectual trends in both philosophy and science in the twentieth century (Stadler, 2015), which was unnaturally facilitated by the rise of Nazism that spurred its members and advocates to emigrate. Many of them eventually landed in the United States. Herbert Feigl was the first to leave in 1931 (after having spent the previous academic year at Harvard on an International Rockefeller Research Fellowship), taking a position at the University of Iowa where he stayed for nine years before accepting a faculty position at the University of Minnesota in 1940. In 1953, thirty years after initiating his graduate work, Feigl founded MCPS, which is still engaged in a similar enterprise of clarifying the basic assumptions, concepts, and methods of the sciences. A central reason for this continuity is a commitment to the ideal of intellectual community.

What is intellectual community? Perhaps it is easiest to begin with what it is not. Intellectual community is not a shared commitment to a set of ideas or a school of thought. Despite common impressions, members of the Vienna Circle disagreed vehemently about many substantive claims: "it teemed with vociferous controversies and silent misgivings" (Sigmund, 2017, p. 7); "discussion was often heated but mostly civil ... what was demanded in any discussion was clarity" (Edmonds, 2020, p. 21). In a self-proclaimed manifesto (Stadler & Uebel, 2012), members noted that their group "is characterized not so much by theses of its own, but rather by its basic attitude, its points of view, its direction of research." Intellectual community is not simply a gathering of academics at a local university or college who interact on a regular basis because of their shared place of work. The Vienna Circle served as a center of gravity such that only some members were fulltime denizens of the city, and many were not formally associated with the university. It also is not the same as a professional society that one can have formal membership in through paid dues and receive an associated peer-reviewed journal or attend annual or biennial conferences. Informal networking, whether in the form of supporting job applications or traveling to connect with like-minded individuals elsewhere—with lots of disciplinary boundary crossing (and criticism)—was a hallmark of the Vienna Circle.

One defining feature of the Vienna Circle's intellectual community was a shared orientation: take seriously the latest developments in logic, mathematics, and the sciences, while avoiding numinous claims that appeared erudite but ultimately lacked meaning. Suspect claims were usually labeled "metaphysics" and included heterogeneous ideas such as Platonic Forms, Hegel's world spirit (*Weltgeist*), and Heidegger's *Dasein*. These were exposed as "devoid of factual meaning" or inherently confused through a systematic, logical analysis of language (Feigl, 1943). However, Vienna

Circle members reserved much of their most strident criticism for each other's work. Another defining feature was regularity in timing and format. Although not everyone was in attendance each time the group met, 10–20 met routinely every other Thursday night. They followed a predictable schedule that included an opening presentation or reading (sometimes by a guest) followed by intense general discussion.

A third feature of intellectual community in the Vienna Circle was its relatively flat social structure, which contained diverse participants from graduate students to full professors without an overly hierarchical deference characterizing their interactions. Although faculty were more likely to talk than graduate students, there were no formal rules against anyone participating in the conversation. Importantly, the Vienna Circle grew out of earlier initiatives and was not established overnight. An initial attempt at formulating a similar discussion group took place more than two decades earlier and included both Hahn and Neurath, though it did not hold together. Intellectual community does not arise instantaneously and typically accumulates momentum over time (and can be thwarted by external factors, such as a world war). This organic growth often involves several anchor members working together to draw others in (e.g., some referred to the group as the "Schlick-Zirkel" early on) and the existence of sister groups that nurture a similar ethos in a different geographic location (e.g., the Society for Empirical Philosophy in Berlin, which began in 1927 and was organized by Hans Reichenbach and Richard von Mises). Finally, intellectual community is interdisciplinary in that its composition is not confined to those sharing a similar training or methodology. The precise composition can be quite variable and change over time, but it does not become a center of gravity for curious scholars if narrowly circumscribed by disciplinary boundaries. Scholarly conversation that sustains intellectual community ranges widely even if there is an overarching topical focus.

Many of these features are present in the research groups ("interest groups") hosted in hybrid format at MCPS. The Center Discussion Group, which is the original group initiated by Feigl, meets biweekly to discuss books in general philosophy of science. The Biological Interest Group meets weekly to discuss a paper oriented to the life sciences, whether conceptual, empirical, historical, or theoretical. Other regular groups focus on early modern natural philosophy, logic and mathematics, physics, and social and behavioral sciences (https://cla.umn.edu/mcps/research-programs/research-groups). Additionally, some groups are organized around a theme, either derived from an existing project or as a prelude to applying for a grant. The Individuality and Modularity Interest Group corresponds to the former category and revolved around an Interdisciplinary Doctoral Fellowship to a graduate student fellow working in evolutionary biology. The Reproducibility Working Group corresponds to the latter and spawned the four-year project, The Many Faces of Reproducibility (https://reproducibility.umn.edu/), whose outputs included public events, symposia, and workshops, as well as a forthcoming volume in the Minnesota Studies in Philosophy of Science (Fletcher and Love, forthcoming).

A crucial ingredient of intellectual community, especially for interdisciplinary groups of disagreeing scholars to engage one other fruitfully, is intellectual hospitality. This involves conversation and requires time. "It's important to have a community of scholars who gather together and talk. If faculty do not practice conversation of the best sort among themselves, it is unlikely that they can bring these skills to bear in the classroom [and elsewhere]" (Cohen, 2023, 162). Faculty, postdocs, and graduate students are always "busy" and focused on their own projects. Motivating them to leave their building and join others for an interdisciplinary conversation can be difficult. However, there is a deep hunger for richer scholarly engagement as the university corporatizes and increasingly adopts a utilitarian orientation to the activities of its members. Although there is not a "good" or "ideal" time to undertake intellectual hospitality for conversations across disagreement and disciplines—other things will always seem more urgent—the same is true for a genuinely hearty meal. Just as a home-cooked dinner can invite sustained and significant interaction among those who consume it in a way that fast food cannot, so also a well-structured intellectual conversation across disciplines will be attractive to those exhausted by endless meetings whose meaning and import is far from clear.

Beyond getting people together, other qualities should mark conversations that exemplify intellectual hospitality. Organizers should focus on topics and themes of interest to those attending and encourage participation through reading one another's work. Meeting times should be as convenient as possible and occur regularly so people can plan their schedules around them without asking. Discussions should revolve around understanding claims and why they are being made, identifying standards of evaluation rather than fixating on positions or schools of thought. Habits of active listening should be cultivated to discern patterns of assumptions and styles of reasoning, especially those that might differ from your own. Participants should be comfortable advancing claims, discussing them vigorously, modifying them accordingly, and sometimes relinquishing them reluctantly. Those leading conversations should watch for the introduction of jargon and do "vocabulary checks" to ensure that everyone is tracking. Sometimes a pause to convey relevant background information that might not be shared by everyone is warranted and pays dividends for engaging more people. Opportunities for everyone to speak, especially near the beginning, help prevent discussion from being dominated by one or two individuals. At the same time, opportunities can be "passed" so that those who feel uncomfortable can accustom themselves to the group dynamic. These "rules of thumb" for interacting are not complicated but do require intentional deployment. Scholars of different disciplinary backgrounds who participate will quickly discern how a culture of intellectual hospitality suffuses conversations.

Why should we care about the ideal of intellectual community at this moment in history? What is so important about the model of scholarly interaction that Feigl imported from Vienna to Minneapolis (and is observable elsewhere)? Having just passed the centenary of the Vienna Circle's origins (1924–2024), there are some interesting parallels between the political upheaval of the 1920s and our current situation. At that time, Vienna was socially turbulent; this segued into unparalleled instability due to the rise of Nazism. Economic turmoil came and went repeatedly during this period. The University of Vienna was embroiled with political infighting and anti-Semitism was increasingly on display, marking daily life and manifesting as riots at the university and elsewhere. Since several of the Circle's members were Jewish, its leaders became targets. "The so-called German Students' Union published blacklists of objectionable professors. These lists included Moritz Schlick … [and] anyone who was suspected of Marxist leanings or of having a Jewish background. Neither criterion applied to Moritz Schlick, but by association, he was seen as guilty" (Sigmund, 2017, p. 173). In 1936, Schlick was murdered by one of his former students (Johann Nelböck) and the individual was subsequently exonerated by the new political regime (Edmonds, 2020). The Vienna Circle had already begun disintegrating under these socio-political pressures, as well as the death of Hahn in 1934. In fact, the situation continued to be difficult (though not nearly as strident) for those emigres with left-leaning politics during the McCarthy period that swept over the United States in the decade after the end of World War II and left an enduring mark on philosophy of science (Reisch, 2005).

Although our current situation has many differences, some will recognize eerie parallels of blacklists, open challenges about the nature, scope, and significance of philosophical (and other forms of academic) investigation, and universities experiencing draconian financial circumstances. The ideal of intellectual community is important to keep in view because it provides the circumstances whereby some of the deepest wrestling with ideas that matter can occur and be propagated. Genuine intellectual community is fecund. For example, a common outcome of intellectual community is publication; the Vienna Circle left its mark everywhere, and not simply by advancing the scholarship of its members. The International Encyclopedia of Unified Science, an unfinished initiative of the Vienna Circle consisting of nineteen monographs published from 1938 to 1969, generated Thomas Kuhn's *The Structure of Scientific Revolutions* (Kuhn, 1962), arguably the most influential historical-sociological study of the sciences in the twentieth century (selling a million copies and cited more than 130,000 times over the past fifty years). Feigl also was involved in the birth of two peer reviewed journals that remain central organs of cutting-edge inquiry on the contemporary landscape of philosophy: *Philosophical Studies* and *Philosophy of Science*. A similar scholarly fecundity is visible in both Egenis (https://www.exeter.ac.uk/research/centres/egenis/outputs/) and MCPS (https://cla.umn.edu/mcps/publications/minnesota-studies-philosophy-science).

To many outside observers, the stereotypical image of a scholar is a lone individual toiling away in the archives or in isolation to achieve maximal concentration—to channel mental exertion toward new ways of looking at the world. The parallel in the sciences is the lone genius who makes fundamental theoretical breakthroughs in sudden moments of insight or through strategic experiments that uncover nature's inner workings. Few are enamored with this picture for contemporary science where journal articles in astronomy and genomics now have hundreds of contributing authors; historians have shown that the image was false of the past as well. However, humanistic scholarship and philosophy (in particular) have retained an individualistic image of lone thinkers accomplishing heroic feats of cognitive strength in isolation. Keeping intellectual community in focus suggests otherwise, as Feigl recognized in contemplating his career. “As I reflect on my motivations in connection with the Minnesota Center for Philosophy of Science, it seems fairly clear to me that my formative experiences in the Vienna circle … encouraged me to endeavor collaborative teamwork in philosophical research” (Feigl, 1969, p. 669). This kind of collaboration requires intentional effort to build and maintain, something that Feigl was famous for and something current resident fellows and affiliates of MCPS and Egenis members strive to achieve.

11.4 Calling Out an Intellectual Legacy

My aim in this chapter has been to pay homage to John Dupre’s scholarly and administrative legacy by drawing attention to ways in which my own efforts follow in the footsteps he has already taken. As highlighted in the introduction, I have ignored individual collaborations between philosophers and scientists or other scholars (of which many could be cited) and focused on institutional level endeavors like interdisciplinary large-scale grant projects and research centers embedded in a university context. These types of endeavors often require substantial interactions with university bureaucracies because they cut against the grain of how the modern university is organized (i.e., disciplinarily), especially its increasing tendency toward quantitative metrics over genuine intellectual engagement and community. The Centre for the Study of Life Sciences and the Minnesota Center for Philosophy of Science are two places where these efforts are exemplified, both in terms of regularly scholarly interaction and larger funded undertakings. Unlike conceptual analysis and the method of cases, these philosophical investigations are carried out collaboratively with scientists, humanists, and other scholars to leverage diverse methods to better understand how the sciences work and identify their genuine (and spurious) implications. They harken back to forms of natural philosophy, when disciplinary boundaries had not yet emerged, and represent an invigorating and productive way forward for philosophical inquiry. I’ve had the honor of being deeply involved in these efforts and count myself fortunate to have followed a similar path to the one John Dupré trailblazed through Egenis at the University of Exeter.

References

Babcock, G., & McShea, D. W. (2021). An externalist teleology. *Synthese, 199*, 8755–8780.

Barnes, B., & Dupré, J. (2008). *Genomes and what to make of them*. University of Chicago Press.

Bertalanffy, L. (1952). *Problems of life: An evaluation of modern biological thought*. John Wiley & Sons Inc.

Bigelow, J., & Pargetter, R. (1987). Functions. *Journal of Philosophy, 54*, 181–196.

Blumberg, A. E., & Feigl, H. (1931). Logical positivism: A new movement in European philosophy. *Journal of Philosophy, 28*, 281–296.

Bolotin, E., Melamed, D., & Livnat, A. (2023). Genes that are used together are more likely to be fused together in evolution by mutational mechanisms: A bioinformatic test of the used-fused hypothesis. *Evolutionary Biology, 50*, 30–55.

Bourrat, P. (2023). A Pricean formalization of Gaia. *Philosophy of Science, 90*, 704–720.

Boyle, R. A., Moody, E. R. R., Babcock, G., McShea, D. W., Álvarez-Carretero, S., Lenton, T. M., & Donoghue, P. C. J. (2025). Persistence selection between simulated biogeochemical cycle variants for their distinct effects on the Earth system. *Proceedings of the National Academy of Sciences USA, 122*, Article e2406344122.

Cohen, P. M. (2023). *Talking cure: An essay on the civilizing power of conversation.* Princeton University Press.

Davies, J., & Levin, M. (2023). Synthetic morphology with agential materials. *Nature Reviews Bioengineering, 1*, 46–59.

de Framond, L., Brumm, H., Thompson, W. I., Drabing, S. M., & Francis, C. D. (2022). The broken-wing display across birds and the conditions for its evolution. *Proceedings of the Royal Society b: Biological Sciences, 289*, 20220058.

Dresow, M., & Love, A. C. (2023). Teleonomy: Revisiting a proposed conceptual replacement for teleology. *Biological Theory, 18*, 101–113.

Dupré, J. (1981). Natural kinds and biological taxa. *The Philosophical Review, 90*, 66–90.

Edmonds, D. (2020). *The murder of Professor Schlick: The rise and fall of the Vienna Circle.* Princeton University Press.

Feigl, H. (1943). Logical empiricism. In D. D. Runes (Ed.), *Twentieth century philosophy: Living schools of thought* (pp. 373–415). Philosophical Library Inc.

Feigl, H. (1969). The Wiener Kreis in America. In D. Fleming & B. Bailyn (Eds.), *The intellectual migration: Europe and America, 1930–1960* (pp. 630–673). The Belknap Press of Harvard University Press.

Fletcher, S. C., & Love, A. C. (Eds.). *Forthcoming. Mapping the frontiers of scientific reproducibility: Foundational questions and intervention strategies.* University of Minnesota Press.

Forber, P. (2020). Contemporary teleology. In J. K. McDonough (Ed.), *Teleology: A history* (pp. 255–278). Oxford University Press.

Ghiselin, M. T. (1994). Darwin's language may seem teleological, but his thinking is another matter. *Biology & Philosophy, 9*, 489–492.

Gotthelf, A. (2012). The elephant's nose: Further reflections on the axiomatic structure of biological explanation in Aristotle. *Teleology, first principles, and scientific method in Aristotle's biology* (pp. 186–196). Oxford University Press.

Gray, A. (1876). Evolutionary teleology. In A. Hunter (Ed.), *Darwiniana: Essays and reviews pertaining to Darwinism* (pp. 293–320). The Belknap Press of Harvard University Press.

Haig, D. (2020). *From Darwin to Derrida: Selfish genes, social selves, and the meanings of life.* MIT Press.

Hill, J., Oderberg, D. S., Gibbins, J. M., & Bojak, I. (2022). Mistake-making: A theoretical framework for generating research questions in biology, with illustrative application to blood clotting. *The Quarterly Review of Biology, 97*, 1–13.

Hill, J., Oderberg, D. S., Austin, C., Cinotti, F., Bojak, I., & Gibbins, J. M. (2026). Mistakes in action: On clarifying the phenomenon of goal-directedness. *Biological Theory*, 21, 37–50.

Huemer, M. (2015). The failure of analysis and the nature of concepts. In C. Daly (Ed.), *The Palgrave handbook of philosophical methods* (pp. 51–76). Palgrave Macmillan.

Kriegman, S., Blackiston, D., Levin, M., & Bongard, J. (2021). Kinematic self-replication in reconfigurable organisms. *Proceedings of the National Academy of Sciences USA, 118*, Article e2112672118.

Kuhn, T. (1962). *The structure of scientific revolutions.* University of Chicago Press.

Laland, K., Odling-Smee, J., & Endler, J. (2017). Niche construction, sources of selection and trait coevolution. *Interface Focus, 7*(5), 20160147.

Lennox, J. G. (1993). Darwin was a teleologist. *Biology & Philosophy, 8*, 409–421.

Livnat, A., & Love, A. C. (2024). Mutation and evolution: Conceptual possibilities. *BioEssays, 46*, 2300025.
Livnat, A., & Melamed, D. (2023). Evolutionary honing in and mutational replacement: How long-term directed mutational responses to specific environmental pressures are possible. *Theory in Biosciences, 142*, 87–105.
Love, A.C. (2005). *Explaining evolutionary innovation and novelty: A historical and philosophical study of biological concepts* (Ph.D. thesis), University of Pittsburgh.
Love, A. C., & Dresow, M. (2022). Organizing interdisciplinary research on purpose. *BioScience, 72*, 321–323.
McLaughlin, K. A., & Levin, M. (2018). Bioelectric signaling in regeneration: Mechanisms of ionic controls of growth and form. *Developmental Biology, 433*, 177–189.
McShea, D. W. (2012). Upper-directed systems: A new approach to teleology in biology. *Biology & Philosophy, 27*, 663–684.
Melamed, D., Nov, Y., Malik, A., Yakass, M. B., Bolotin, E., Shemer, R., Hiadzi, E. K., Skorecki, K. L., & Livnat, A. (2022). *De novo* mutation rates at the single-mutation resolution in a human HBB gene region associated with adaptation and genetic disease. *Genome Research, 32*, 488–498.
Melamed, D., Shemer, R., Bolotin, E., Yakass, M. B., Fink-Barkai, D., Hiadzi, E. K., Skorecki, K. L., & Livnat, A. (2025). *De novo* rates of a *Trypanosoma*-resistant mutation in two human populations. *Proceedings of the National Academy of Sciemces USA, 122*, e2424538122.
Moreno, A., & Mossio, M. (2015). *Biological autonomy: A philosophical and theoretical enquiry.* Springer.
Moreno, A. & Peretó, J. (2026). *An evolutionary story of agency: How life evolved to act on its own.* Springer-Nature.
Mossio, M., & Bich, L. (2017). What makes biological organisation teleological? *Synthese, 194*, 1089–1114.
Neander, K. (1991). Functions as selected effects: The conceptual analyst's defense. *Philosophy of Science, 58*, 168–184.
Newman, S. A. (2023). Inherency and agency in the origin and evolution of biological functions. *Biological Journal of the Linnean Society, 139*, 487–502.
Odenbaugh, J. (2019). Functions in ecosystem ecology: A defense of the systemic capacity account. *Philosophical Topics, 47*, 167–180.
Okasha, S. (2018). *Agents and goals in evolution.* Oxford University Press.
Perlman, M. (2004). The modern philosophical resurrection of teleology. *The Monist, 87*, 3–51.
Reisch, G. A. (2005). *How the Cold War transformed philosophy of science: To the icy slopes of logic.* Cambridge University Press.
Richards, R. J. (2009). Darwin's place in the history of thought: A re-evaluation. *Proceedings of the National Academy of Sciences USA, 106*, 10056–10060.
Riskin, J. (2016). *The restless clock: A history of the centuries-long argument over what makes living things tick.* University of Chicago Press.
Rosenblueth, A., Wiener, N., & Bigelow, J. (1943). Behavior, purpose and teleology. *Philosophy of Science, 10*, 18–24.
Rosslenbroich, B. (2023). *Properties of life: Toward a theory of organismic biology.* MIT Press.
Ruse, M. (2018). *On purpose.* Princeton University Press.
Schroeter, L. (2004). The limits of conceptual analysis. *Pacific Philosophical Quarterly, 85*, 425–453.
Sfara, E., & El-Hani, C. N. (2023). Ecosystem health and malfunctions: An organisational perspective. *Biology & Philosophy, 38*, 37.
Sigmund, K. (2017). *Exact thinking in demented times: The Vienna Circle and the epic quest for the foundations of science.* Basic Books.
Simpson, G. G. (1958). The study of evolution: Methods and present status of theory. In A. Roe & G. G. Simpson (Eds.), *Behavior and evolution* (pp. 7–26). Yale University Press.
Stadler, F. (2015). *The Vienna Circle. Studies in the origins, development and influence of Logical Empiricism.* Springer.

Stadler, F., & Uebel, T. (Eds.). (2012). *Wissenschaftliche Weltauffassung. Der Wiener Kreis.* Springer.

Strevens, M. (2019). *Thinking off your feet: How empirical psychology vindicates armchair philosophy.* Harvard University Press.

Walsh, D. M. (2008). Teleology. In M. Ruse (Ed.), *The Oxford handbook of philosophy of biology* (pp. 113–137). Oxford University Press.

Walsh, D. M. (2015). *Organisms, agency, and evolution.* Cambridge University Press.

Walsh, D. M., & Rupik, G. (2023). The agential perspective: Countermapping the modern synthesis. *Evolution & Development, 25*, 335–352.

Watson, R. (2024). Agency, goal-directed behavior, and part-whole relationships in biological systems. *Biological Theory, 19*, 22–36.

Wimsatt, W. C. (1972). Teleology and the logical structure of function statements. *Studies in History and Philosophy of Science, 3*, 1–80.

Wright, L. (1973). Functions. *The Philosophical Review, 82*, 139–168.

Zammito, J. (2018). *The gestation of German biology: Philosophy and physiology from Stahl to Schelling.* University of Chicago Press.

Part IV
Concluding Essays

Chapter 12
The Mighty Worlds We Half Create

Phillip Kitcher

For John Dupré

Abstract John Dupré challenged us long ago to think of "promiscuous realism" and of "the disorder of things". He argued cogently against any grand programme of unified science, and for the thought that there are many cognitively valuable ways to divide up the world. His arguments have convinced me, and I have retreated from thinking about explanation as achieved through unification to the view that there is no general theory of explanation, or even of scientific explanation. But John has changed too, or, perhaps, his metaphysical tendencies have demanded their full expression. He has started to think in terms of a unified ontology for the world, one that takes processes as its fundamental constituents. The shift does not retract his opposition to reductionism, but reflects his conviction that our understanding of many phenomena, particularly those connected with the living world, is best advanced by downplaying things and prioritizing processes. I find the idea of sometimes thinking in terms of processes highly congenial, but I am ever more convinced of the value of different ways of conceiving things. My lecture will explore a tradition of pluralist (promiscuous?) constructivism that runs through Emerson, William James, and Dewey to Kuhn and Nelson Goodman. It is grounded in the thought of a world of experience that can be shaped and reshaped to facilitate our purposes, not only our inquiries but also our interactions with one another. The values to which we may legitimately appeal in our reshapings need not be restricted to the purely cognitive. This tradition can be thought of as a version of Kantianism with moveable categories. It clearly emerged in the writings of some post-Kantians. Its roots, though, as I'll suggest, are slightly earlier, in a different source.

P. Kitcher (✉)
Columbia University, New York, NY, USA
e-mail: psk16@columbia.edu

S. Leonelli et al. (eds.), *Life Beyond Human Boundaries: John Dupré's Philosophy and the Interdisciplinary Study of the Life Sciences*,
https://doi.org/10.1007/978-3-032-26977-5_12

In the beginning, there was philosophical darkness on the face of Exeter. And the Solons of the University said "Let there be John Dupré". And Lo! There was Egenis; and a Philosophy Department. And the Solons saw that it was good.

A lot has happened here since this university courted Regenia Gagnier as a worthy successor to the redoubtable Michael Wood, and wondered what should be done for—or with?—her husband. John and Regenia had occupied tenured positions in two outstanding departments at a major American university. They left partly because of their dislike of the right-wing miasma that hangs over some segments of Stanford, and partly because they wanted their boys, Gabe and Julian, to grow up in a different environment. John arrived as a part-time Senior Research Fellow in the Exeter English Department, with another part-time job at Birkbeck (in London). He was the only philosopher around, and, at the beginning, nobody had the bright idea of creating a philosophy department around him.

So, like other Creation myths, the version with which I began is badly inaccurate. Nobody foresaw what John might be able to do for this university. But then, nobody here knew him.

I first met him forty-odd years ago. It was at an APA meeting. On that we agree. We disagree, though, about how it happened. In John's version, I went over to talk to him; in mine, he came over to talk to me. The difference is significant. Each of us credits the other for starting a beautiful friendship.

Some of John's personal qualities are well known to all of us. He is famous around the philosophy profession for his wit and social sparkle. More importantly, he is much loved for the kindness that is second nature to him. He has helped numerous philosophers find their feet. He has given himself to promote valuable ventures. Egenis, for example, began as the lowliest of the funded centers set up to explore the social implications of genomic research. It established itself as the principal British institution engaged in that project. All because of John. I recall a conversation with Nick Rose, when we were fellow members of the advisory board in the center's early days. We agreed that the board was unnecessary. John knew what he was doing, and our best response was simply to clap.

Yet there is something even more important to know about John. He has a true gift for friendship. That, I fear, is increasingly rare in a world where "friending" happens casually. But, like Regenia, John knows how to talk and how to listen, how to give and how to receive. He knows what real friendship is. He has been not only a close philosophical interlocutor, but also a cherished friend. So it is a great honor and pleasure to talk about him today.

*

But it is time to stop embarrassing him, and turn to philosophy. John is one of the most eminent philosophers of science in the world. The Disorder of Things has become a classic. His books exposing the ways in which biology and economics are adapted to buttress socially obnoxious conclusions are rightly influential. He has expanded our views of what biology is, and how its central themes should be conceived. All this is commonplace.

Today, I want to juxtapose John Early and John Late. At the time we first met, John had just published a brilliant article in which he introduced—and defended—what he called "promiscuous realism", the thesis that Nature is multiply jointed, with real divisions that answer to different explorations. More recently, in his Gifford Lectures, he has let his metaphysical proclivities hang out (as the Nixon gang might say), suggesting that we should think of processes as the fundamental entities in the world. Despite his enthusiasm for disorder, expressed in a fierce opposition to reduction in the sciences, he wants a unified ontology, one based on processes. How does the early promiscuity fit with later monogamy?

Over the four decades of our friendship, we seem to have traveled in opposite directions. In the early 1980s, while I was one of the founding members of the Unification Church, believing that scientific explanations unified the phenomena, John was skeptical of such proposals. Since then, I have come to think the quest for a general theory of explanation misguided, viewing scientific explanations as a diverse mixture. The pluralism I espoused about species concepts, and later about units of selection, has been extended more widely. Meanwhile, John has campaigned for a unified ontology, in which processes are fundamental.

In his Gifford Lectures, he makes a strong case for viewing some biological phenomena in terms of processes. He argues for the fruitfulness of regarding organisms and species as processes rather than as things. I agree that using processual descriptions to reconceptualize some biological issues brings certain features into higher relief, thus facilitating particular kinds of explanations, predictions and interventions. But I have two questions about the metaphysical ambition to replace an ontology of things with an ontology of processes. The first, and less important, stems from a suspicion that, without some reference to things there will be no available points of stability from which we can identify and individuate processes. The second is generated by my latter-day promiscuous impulses. Should we assume that a single ontological picture will be well-suited to all our purposes? Perhaps, in the spirit of John Early, we might require different ways of characterizing the world as we move from project to project.

The remainder of this talk will mostly be devoted to pursuing this second issue. First, though, a quick look at some points about individuation.

*

Consider an uncontroversial instance of a process. At 10 a.m. one Saturday, John goes out to weed the garden. Two hours later, he takes a break for lunch. From 2 p.m. to 5 p.m. the weeding continues, after which he comes in for a drink and checks on the test match. The process of John's weeding falls into two non-continuous segments. How do we identify it?

Normally, we do that by starting with things: John, his garden, weeds, Moor House, Regenia's specialty sandwiches, the garden again, more weeds. All of these could, of course, be viewed as processes. That would raise questions about how to identify them.

One of these things seems especially relevant to our initial attempt at identifying the Saturday weeding. John. Process-John starts when two other processes combine; he retains a special connection with one of them for a few months; after that he is detached, his worldline no longer continuous with either of the generating processes, although, for a significant period, often in proximity to one or both of them.

Various properties distinguish Process-John from lots of processes with geometrically similar worldlines. Are there enough to differentiate Process-John from all these other processes?

The most promising way of addressing that question is, I think, to invoke a spatio-temporal coordinate system. Pick an origin (the Big Bang?); set up three spatial axes, and one temporal axis; then we can identify the Earth-process as originating in a particular region of this four-dimensional structure, locate the Britain-process, and finally, Process-John's point of origin. Once we have that, Process-John has been identified.

This will solve the problem without invoking any material objects. They can all be regenerated as processes originating in particular space-time regions, and thereafter grouped by shared common properties, or by their historical connections to one another. But we still have to invoke some thingy entities, some fixtures in the Heraclitean flux. We need space-time points, an origin, and axes.

I hope you have been impatient with my tangled metaphysics. I've introduced it only as a stepping-stone to a more important point. Lots of everyday talk about the world and investigations of it would be inconvenient to the point of unfeasibility if it had to be conducted by supposing that it's processes all the way down. Our normal practices of conversation and inquiry individuate processes by appealing to things. For some purposes, it is valuable to treat some of those things as processes. For different purposes, it is valuable to view them as stable objects.

John could conduct his weekend weeding by using a taxonomy supplied by botany. Glancing down at his lawn, he says "Ah! Bloody Taraxacum officinale, again! Root it out!". A few moments later, he smiles benignly to observe another member of the Asteraceae, and leaves a fine specimen of Bellis perennis to grow. At the end of the day, he digs up an Allium cepa for dinner, allowing his Allium millennium to continue to flower. But actual John doesn't say any of these things: he counts the dandelion as a weed, the daisy and the lily as flowers, and the onion as a vegetable. That makes his life far easier than having in his head a long list of Latin names of plants he is out to annihilate, a similarly long list of those he aims to cultivate, and a somewhat shorter list of those with which the family and lucky guests will be fed.

John Early saw this very clearly in that brilliant article. Different vocabularies suit different human purposes. That was the core insight of his promiscuous realism. Why then insist on a single ontological category as suited to all talk and all inquiry? Why even suppose that a single vocabulary, process-talk, works for all biological purposes?

Are we witnessing the Rake's Progress? Has John Early become a fussy Puritan?

*

Rebutting the charge and achieving reconciliation proves relatively easy. Instead of claiming that processes are ontologically fundamental, and that the world consists of processes, John could offer a more modest thesis. Human beings have many different purposes, and to expedite all these purposes different ways of dividing up reality are needed. No classificatory system is the equivalent of the ideal Swiss Army Knife. Processes are often a good idea, especially in parts of biology. But any division that is workable for some cluster of human purposes is natural. Hence there are many overlapping systems of natural kinds. Gardeners, cooks, and biological systematists all pick out real divisions in nature.

This is the position I think the author of The Disorder of Things ought to adopt—although he might well want to refine my formulation of it. In the remainder of this talk, I want to contrast it with a slightly different view, one that retains the promiscuity but dilutes the realism. The motivation for this alternative view comes from suspicion aroused by one of the phrases I have employed. What exactly are "real divisions in nature"? Are there any such things?

Let's move first from the garden to the kitchen. Julian has come home for a visit, and prepares the family dinner. It is to be a vegetable extravaganza. He will use asparagus, artichokes, beets, beans, daikon, eggplant, fennel, and more. No celery, though, since one of the diners is allergic to that. Despite the allergy, celery still counts as a vegetable. Why? Because most people can digest it.

Some years pass. As a result of new techniques in preparing processed food, consumed voraciously by children around the world, most people come to be allergic to celery. It is no longer sold in grocery stores, and is scratched from the canonical list of vegetables. There's compensation, though. Potato roots and eggplant leaves are now digestible by a majority of human beings. Supermarkets advertise them as gourmet treats.

Which of these classes of plant parts picks out a real division in nature: the old one? the new one? both?

A second example. Britain is an island. Which chunk of the Earth's surface constitutes it? To answer the question, we need to fix the island's perimeter. But now John's promiscuity surges back. Is it the land exposed at the lowest tides, so that the sea invades a part of Britain every day? Or at the highest tide, allowing people to venture into what properly belongs to the sea, and still remain dry, provided they choose the right time of day? Or somewhere in between? These questions will be answered differently by geologists, kayakers, hikers, and commissions concerned with protecting against storm surges and floods. Different human purposes, different perimeters.

Are they all real divisions in nature?

For a third case, I shall return to a question John and I debated fifteen or so years ago. Are there human races? From the 1960s on, a significant number of thinkers, many of them anthropologists, recommended a negative answer. They hoped to eliminate racism by debunking the concept of race. A worthy aim, although whether you can achieve it by undermining a concept strikes me as dubious. John, a firm eliminativist, was similarly worried.

Human history has been dominated by ugly conceptions of race. They have often supposed a hierarchy of races. They have usually marked some groups as essentially and inevitably inferior. It is sobering to read Jefferson's Notes on the State of Virginia, and to discover that a man, enlightened in many respects, could hold such repugnant views about a group of people, including one woman with whom he had sexual relations.

Yet vile as these social and folk concepts of race have undoubtedly been, there is an alternative notion, articulated by biologists who have freed it from any essentialist or hierarchical assumptions. Contributors to the evolutionary synthesis of the 1930s, most prominently Theodosius Dobzhansky, were interested in the internal structure of biological species and the existence of subgroups that might exhibit the beginnings of speciation—species in statu nascendi, as they called them. That tradition was continued by L. L Cavalli-Sforza and his younger colleague, Marcus Feldman, at Stanford, who investigated the human population (Rosenberg et al. 2002). Both were determinedly anti-racist, and Feldman, in particular, wrote brilliantly about the flaws in alleged demonstrations of racial differences in socially significant properties.

One of the major scientific articles published in this century is a paper by Feldman and his research team. Using genomic data on the frequency of genetically meaningless sequences among various human populations, they identified various "clusters" within Homo sapiens. They were thus able to answer a sequence of questions. If you divided our species in two on the basis of genetic similarity, where would the partition come? Suppose you made the division threefold, what would it be? And so on, for any number you care to choose.

Here are the answers for the first few questions. Two divisions: America and Africa (including Eurasia). Three divisions: America, Africa, and Eurasia. Four divisions: East Asia is hived off. Five: separate out Oceania. The first surprise comes at six: a small population in Pakistan is differentiated.

Feldman and his colleagues were cautious, punctilious in referring to the subgroups as "clusters". They hypothesized, in ways motivated by the biological species concept, that the divisions represented the history of human migrations, and the relative strength of the barriers to mating between groups of people. Almost immediately, though, journalists imposed a different language. Writing in the New York Times, Nicholas Wade advertised the article as identifying human races.

Since then, a considerable amount of sophisticated philosophical work has been done to articulate a non-racialist approach to human races building on the science I have sketched. Apparently, dividing our species into something like races is useful for demographic research. With appropriate circumspection, this concept can be deployed without any of the hideous assumptions about racial differences that have done so much damage in the past. Nevertheless, whether or not the word 'race' is used to denote the clusters, it is entirely reasonable to worry that circumspection will lapse, that old associations will leak in, and that groundbreaking scientific work will be distorted to support socially destructive conclusions. We are dealing with a concept that straddles two domains—science and society. Should we ban it, despite its demographic utility, as eliminativists would urge? Or should it be retained, surrounded with some precautions for avoiding its social abuse? These are ethical questions, and different people offer different answers. Tommie Shelby, for example, has argued for retaining the concept of race on the grounds that it is useful for generating a form of solidarity that will help undo the harms of past racialism.

I draw two morals from this example. First: off-the-cuff answers are premature; serious ethical inquiry is required, and that can only proceed by inclusive deliberation, well-informed and conducted to engage with all diverse perspectives. Second, the constructive—creative!—aspect of human boundary-drawing shows up very clearly in this example; if there are races, how many are there? Feldman's work seems to answer that with another question—"How many do you want?"

*

The three examples I have offered could be read as revelations of the promiscuity John Early celebrated. Promiscuous realism revels in an unexpectedly vast number of natural divisions in reality. Yet many of those divisions seem to arise from human needs, human desires, and to be properly settled by our ethical explorations. Are we tracing the independent structure of the world or drawing boundaries to suit ourselves? Is "the trail of the human serpent" over everything?

William James thought so. The idea of the organization of the world as humanly constructed is not only articulated in his Pragmatism lectures (to which I have just alluded) but present already in his Principles of Psychology. Towards the close of the "Stream of Thought" chapter of that seminal work, James raises a pertinent philosophical question.

But what are things? Nothing, as we shall abundantly see, but special groups of sensible qualities, which happen practically or aesthetically to interest us, to which we therefore give substantive names, and which we exalt to this exclusive status of independence and dignity. But in itself, apart from my interest, a particular dust-wreath on a windy day is just as much of an individual thing, and just as much or as little deserves an individual name, as my own body does. (James 1890, 274).

Set on one side the misleading restriction to sensible qualities and the attributions of independence and dignity. The core of James's position is that ontological decisions depend on us and vary with our interests.

This point is developed later in the chapter, by using a suggestive but difficult analogy.

The mind, in short, works on the data it receives very much as a sculptor works on a block of stone. In a sense the statue stood there from eternity. But there were a thousand different ones beside it, and the sculptor alone is to thank for having extricated this one from the rest. … the world we feel and live in will be that which our ancestors and we, by slowly cumulative strokes of choice, have extricated out of this, like sculptors, by simply rejecting certain portions of the given stuff. Other sculptors, other statues from the same stone! Other minds, other worlds from the same monotonous and inexpressive chaos! My world is but one in a million alike embedded, alike real to those who may abstract them. How different must be the worlds in the consciousness of ant, cuttle-fish, or crab! (James 1890, 277).

The final sentence draws on physiological understandings. Recognizing the differences in sensory faculties, the diversity of organismic needs, and the variety in capacities for satisfying those needs, James suggests that different animals must organize their experience in radically different ways. If non-human animals were to speak to us, their ways of identifying objects, kinds of objects and properties would diverge so much from ours that—just as Wittgenstein would later predict—we would not understand them. James's remarks about the cumulative work of our ancestors, however, deny that biological differences are the sole source for variations in world-making. Cultural changes modify human predilections, forcing reorganization. They thus change the world—or at least "the world we feel and live in".

James's analogy is seductive. A promiscuous realist, however, is likely to counsel resisting its charms. Is James arguing that we are completely free in organizing our world of experience (henceforth: Lebenswelt)? The thought of making up the Lebenswelt however we like is belied by our experience. Sometimes the world is a stern Popperian, resisting our efforts to impose a particular kind of organization on it. Realists must amend the analogy. The block of stone cannot be treated as homogeneous, allowing for the liberation of any internal chunk to stand forth as a statue once the surrounding material has fallen away. If the sculptor tries to turn the chisel down some—indeed most—lines, it will slide, moving along neighboring fault lines in the stone. These fault lines represent the structure within the stone. They determine the "natural divisions in reality". They allow a large variety of statues to be built. Many of these correspond to Lebenswelts human beings could never live in, to the worlds of non-human animals, crabs, cuttlefish and the like. Others represent the many worlds inhabitable by human beings who will never develop the interests to make them appropriate. James's constructivist ontology misreads the lessons of promiscuity. Human beings are not the authors—or sculptors—of the Lebenswelt in which they live. They have selected one Lebenswelt from the many, attending to one portion of the structure of reality.

Promiscuous realists thus recognize lots of deep divisions in nature. Animals, including human beings, latch on to some particular subset of these (presumably a small one), to generate the Lebenswelt manifested in their actions. Moreover, as human culture evolves, our species reconstructs the Lebenswelt. We organize the world differently. We might even say something like the following:

At the very least, as a result of discovering oxygen, Lavoisier saw nature differently. And, in the absence of some recourse to that hypothetical fixed nature that he "saw differently," the principle of economy will urge us to say that after discovering oxygen Lavoisier lived in a different world. (Kuhn 1962, 117).

Those sentences come from the tenth chapter—the notorious "X-rated Section X" of *The Structure of Scientific Revolutions.* A section, interestingly, in which Kuhn appeals both to William James and Nelson Goodman, whose later book, Ways of Worldmaking, will recapitulate some of its themes and provoke similar cries of outrage.

But what exactly is this "principle of economy" to which Kuhn appeals? I reconstruct Kuhn's thinking as follows. Our language refers to the constituents of an organized world, our Lebenswelt, about which we can pose, sensibly, questions about the relations among those constituents, about which (if any) are most fundamental, about how many there are, and so forth. Those questions presuppose prior decisions, possibly taken quite unconsciously at the direction of our biological impulses, about how to answer the question James poses: What is a thing? Moreover, in reflecting on language and its relation to the world, we are already presupposing the same scheme of organization. All our discourse is about the Lebenswelt. As Kuhn sees it, the further thought of a relation between language and the underlying structure of the world is an irrelevant extra. It tries to say the impossible: to allege a fit between the organized world and Reality. But the phrases we use can only be metaphors and analogies, picturing Reality in the only language we have, the language of the

Lebenswelt. We project characteristics we recognize in the Lebenswelt onto something beyond it, conceiving The World as a multi-jointed beast or a block of marble with fault lines.

The right strategy is to give up the attempt. Inquiry generally is an attempt to find ways of organizing our experience, selecting those that work to advance our projects, rejecting the many in which we experience resistance. Or, more exactly, with the racial example in mind: we try to find divisions and forms of organization that help in those projects endorsed by ethical inquiry. The structure of the Lebenswelt emerges from the interplay between factual and ethical investigations.

Resistance is all too familiar. What James and Kuhn recommend is avoiding trying to provide a deep metaphysical explanation of the source of resistance. We can say, with John Late, that thinking in terms of things rather than processes gets in the way of our efforts to understand particular facets of evolution. What we should not say is that particular divisions—into biological species, say, or into processes—correspond to the deep structure of the world, whereas others do not, and that the mismatch between some categories and the deep structure is the underlying cause of the resistance we feel. Causal talk relates elements of the Lebenswelt. It does not fathom the relation between the Lebenswelt and some incoherently hypothesized source of resistance.

This is, I think, a deep theme in pragmatism. It reflects James's interpretation of Peirce's principle: we ought to abandon questions that lack implications for human conduct. It is present in his apparently astonishing suggestion that "Truth happens to an idea." It is further developed in Dewey's equally surprising claim that human beings reconstruct the world. Kuhn and Goodman, the latter already in The Structure of Appearance and Fact, Fiction and Forecast (1951), are heirs to this tradition.

As I'm sure many of you already appreciate, the tradition doesn't begin with the classical American pragmatists. We shall come to that. First, however, it's worth seeing how the concerns I have raised arise with especial severity when realism goes promiscuous.

*

Here's another way to present John Early's great insight: Nature doesn't set an agenda for investigators. It isn't divided into units ready to call out to us: "Come over here! I'm important. Investigate me!" The questions we pose arise from our diverse needs and interests. Hence, gardeners and cooks require categories different from those deployed by botanists and evolutionary theorists.

According to George Washington Carver, a weed is a flower growing in the wrong place. Similarly, a vegetable is a (non-flowering) part of a plant human beings can digest. Tastes in flowers change; the human digestive system can be affected by diet, or by other environmental causes. It is easy to envisage Julian having a different array of vegetables for some future extravaganza, or John coming to direct his weeding towards a different collection of plants. John Early can make sense of these changes, by supposing that both categories represent real divisions in nature. There are lots of structures, lots of fault lines, from which people select the ones that are valuable for their purposes. At different times the selections may go differently.

Similarly, reality allows many different perimeters for Britain. Swimmers, hikers, kayakers, geologists, flood engineers can be permitted their different choices: all pick out objective boundaries in the natural world.

How far can this go? Is any continuous curve within the space between the high-tide and the low-tide perimeters sanctioned as natural? Surely there are only a few such curves that are suited to actual human practices. Again, though, that could change. The kayaking boundary might shift with modifications of technology, enabling kayaks to come closer. Or, more fancifully, the boundary might be drawn to suit some new type of activity, involving contact with water or land at what now would appear as a bizarre sequence of places—an activity that comes to dominate the coastal scene. Moreover, the boundaries of Britain are easily contracted or expanded. Parts of the land are ground down, or there are new structures extended as far as you like—to France, Iceland, or the North American shores, say. Does Paignton Pier already count as part of Britain? Naturally?

The emerging concern is the possibility of conceiving any boundary as natural, whether it is used to demarcate the contours of an object, to mark the worldline of a process, or to group objects into some category. At first sight, some hypothetical objects appear absurd. Consider the chunk of space-time made up from Queen Victoria's left big toe, the manuscript of Finnegans Wake, and the largest and most stubborn root among all the dandelions in John's garden. Can we conceive of a human pursuit facilitated by thinking of that as a single object? My treatment of the perimeter of Britain points the way. If we can imagine a coastal activity pursued by a trajectory tracing any arbitrarily assigned curve, we can also imagine a practice—a religion, say—that identifies what we would view as three independent objects as aspects of a sacred trinity.

It looks, then, as if anything goes. To accommodate the full range of potential human pursuits, we must allow any boundary whatsoever to count as a natural division in nature. Promiscuous realism becomes California naturalism. Its slogan is: Whatever. The distinction between selecting from the large menu reality presents, and constructing the categories that work seems to evaporate. Pragmatists will see no difference in conduct, and will suggest abandoning talk of natural divisions.

How could John limit the process of expansion, in which more and more boundaries are certified as natural? By restricting the human purposes to those espoused now? But why should this particular moment in our history be privileged? My third, racial, case offers a more promising suggestion. Don't allow any old purpose to be approved. Just focus on those that could be validated by ethical inquiry. If you accept the ethical permissibility of the Feldman notions of race, that would allow natural divisions to appear for a sequence of choices (how many races do you want?). It would debar, however, those proceeding by appealing to a number of properties traditionally associated with racial divisions.

I doubt that John would wish to pursue this option. For its ability to block alternative ways of demarcation that would generate the same overabundance of natural categories, by deploying physical properties as criteria of demarcation rather than ethically obnoxious ones, is far from clear. Moreover, even if it were successful, it would tip promiscuous realism in an odd direction. The natural divisions of reality would reflect a subset of potential human interests: those according with the results of ethical inquiry. In effect, the natural divisions of reality would represent the ethical structure of the world.

*

John Late should, I think, give up the Nixonian hang-out route of heavy-duty metaphysics. He should become a promiscuous—or, perhaps more decorously, a pluralist—pragmatist. Let me briefly sketch how this might go.

Human inquiry is motivated by our practical need to solve problems, although in the course of our history, such inquiry sometimes generates some new problems arising out of people's curiosity. The questions we pose already embody ways of organizing our experience. But, as we proceed, we discover improvements in how we frame things. Those improvements better serve our purposes. Moreover, the purposes themselves evolve. Sometimes that occurs as the result of processes modifying our views about what is ethically correct. Egenis pursues the implications of genomics without lapsing into discredited forms of eugenics.

The primary function of concepts of truth, accuracy, and correctness is to mark judgments (and other representative vehicles) as suitable for use in further practice and further investigation. It is perfectly reasonable to think of truth as relating some judgments to objects and properties in the Lebenswelt via the familiar Tarskian machinery. Inquiry generally is driven by our attempts to say what is true—or, better, true enough to advance our purposes—within the language we currently speak, and also to amend that language to make our efforts to pursue our aims go more easily and more successfully. Since we have plenty of different legitimate purposes, we often need many different ways of dividing up the same domain of experience. The complete Lebenswelt contains many world-versions among which we move without strain.

People of a metaphysical bent are easily led to ask extra questions when they think about conceptual change and conceptual pluralism. They start from a commonplace observation. Sometimes, as we discover, the language we've been using doesn't work well. We encounter resistance. If you have a metaphysical itch, you want to know why this happens. There are perfectly straightforward answers to specific why-questions like this, given by scientists and historians of science. We can account for the failures of the language of chemistry used before Lavoisier or the difficulties of biological taxonomy if you don't know about evolution. The language of the Lebenswelt suffices for understanding the difficulties of prior versions of the Lebenswelt. Metaphysical itches survive, however. What's wanted is a general account of the source of resistance. Thus arise stories about divisions in nature and analogies about fault lines in blocks of marble. When those stories are pursued further, even a stalwart anti-reductionist may be seduced into talking about "fundamental ontology".

Earlier, I heard resistance in a Popperian voice, and it is useful to extend that analogy. The metaphysical debate with which I have been concerned centers on the limits of human freedom in organizing the Lebenswelt. Which bits come from us, and which are dictated by something beyond us? Within the Lebenswelt we can separate the creative artist from the faithful reporter. If resistance were not a Popperian, if the voice came with three responses—"No!" "Yes!" and "OK"—we could differentiate the places in which we overstepped or conformed to nature's own boundaries from those in which we could draw lines of our own choosing. Unfortunately for the metaphysician, resistance has only two modes: it is either present or absent.

It is no great loss. For we can manage without an answer to the general question of how free we are. We can simply get on with the business of inquiry. Pragmatists will see an alleged difference that makes no difference.

Why, then, isn't John Late a pluralist pragmatist?

*

A conjecture. He views the philosophical approach I have outlined as having a disreputable pedigree. He does not want to go over to the Dark Side.

James and Dewey were profoundly influenced by Ralph Waldo Emerson, hailed in his time as America's greatest thinker. Today, most philosophers would exclude Emerson from the canon. His writings are epigrammatic, loose, couched in the idiom of the pulpit rather than the philosophy seminar. Moreover, he seems to make wild and outrageous suggestions. Like the last sentence of Nature, his first book and, according to some critics his best: "Build therefore your own world!"

Behind Emerson stands the murky tradition of post-Kantian Idealism, a poison injected into the bloodstream of American thought by a British drug addict, a poet, like Emerson, known in his own time as something of a philosopher. This poet, especially enamored of the writings of Schelling, made friends, in the early 1790s, with a fellow-poet, whom he later credited with having written the "first philosophic poem in the English language". In that friendship, I hope, lies the key to winning John's heart for pluralist pragmatism. For the poet who received praise for his philosophic efforts gives the best general statement of the position I know, and does so in poems that John and I both admire.

He writes, for example, lines like these:

> Therefore am I still
> A lover of the meadows and the woods,
> And mountains; and of all that we behold
> From this green earth; of all the mighty world
> Of eye and ear, both what they half-create,
> And what perceive; well pleased to recognize
> In nature and the language of the sense,
> The anchor of my purest thoughts, the nurse
> The guide, the guardian of my heart, and soul
> Of all my moral being.

This passage, in the early poem prompted by his return to Tintern Abbey, is only the beginning of many others that, collectively, illuminate the kind of pragmatist pluralism I have been trying to motivate. I hope our shared love of Wordsworth will help bring John over to the Dark Side.

But, even if, as Wordsworth sees, we often can't tell what we perceive from what we half-create, there are some clear cases. There can be no doubt about John's creative success both in his impact on philosophy and in his institution-building at Exeter. So let us applaud him.

References

Goodman, N. (1951). *The structure of appearance.* Harvard University Press.
James, W. (1890). *The principles of psychology, in two volumes.* Henry Holt and Company.
Kuhn, T. (1962). *The structure of scientific revolutions.* University of Chicago Press.
Rosenberg, N. A., Pritchard, J. K., Weber, J. L., Cann, H. M., Kidd, K. K., Zhivotovsky, L. A., & Feldman, M. W. (2002). Genetic structure of human populations. *Science, 298*(5602), 2381–2385. https://doi.org/10.1126/science.1078351

Chapter 13
Some Final Thoughts on This Volume

John Dupré

Abstract In this chapter, I comment briefly on the earlier chapters in the book. I begin with a discussion of symbiosis, a topic that was foundational to my interest in process philosophy, and that has brought my recent work into closest connection with practicing scientists. I then consider the chapters that reflect my longstanding interest in the relevance of science to understanding the human. Some responses to the challenges to my views on the connection between science and metaphysics are then presented with a little more philosophical detail. The chapter concludes with some thoughts about interdisciplinarity and, finally, the way in which the book itself so nicely exemplifies this ideal.

I must begin by thanking all the contributors and, especially, the editors of this volume. I know how much work and time goes into a project of this kind, and it is truly appreciated. To have so many distinguished colleagues and friends contribute to a book with my name in the title is a great honour and privilege. Most special thanks of all to Sabina Leonelli, who as well as being one of the editors was the driving force behind the conference at which many of these chapters were first presented. The introduction, as well as saying some very generous things about me personally, provided an account of my philosophical work that fitted it into a coherent narrative that I would have struggled to produce myself. Perhaps there are elements of rational reconstruction, but that isn't a bad thing, and it enabled me to understand my own work better.

The chapters of this book touch my philosophical interests from many directions, and I would like to say something about all of them, though I shall order my remarks somewhat differently from the editors' order of the chapters.

J. Dupré (✉)
University of Exeter, Exeter, UK
e-mail: J.A.Dupre@exeter.ac.uk

S. Leonelli et al. (eds.), *Life Beyond Human Boundaries: John Dupré's Philosophy and the Interdisciplinary Study of the Life Sciences*,
https://doi.org/10.1007/978-3-032-26977-5_13

13.1 Process Biology, Symbiosis and Microbiology

Much though philosophers of biology would wish it, most biologists do not take much of an interest in their reflections. However, there have for a while been a few of the best who have engaged deeply and seriously with the philosophy and history of their subject, and I am honoured that two of the best of these, Ford Doolittle and Scott Gilbert, have contributed to this volume. Both scientists address the topic of symbiosis. Gilbert addresses directly the nature of a holobiont, the whole consisting of a multicellular organism and its associated symbiotic microbes, and explains why this must be understood as a process, and Doolittle considers the implications of symbiosis for evolutionary theory. Symbiosis played a central role in my own journey to processualism.

This convergence on topics is no accident. Thomas Pradeu, in his chapter, explicitly proposes that "symbiosis, together with lateral gene transfer, has been for Dupré a major argument and motivation for introducing process ontology". I'm happy to confirm, at least introspectively, that this is true. Indeed, he also offers some bibliographic evidence that this is the element of my work that has received most notice in the scientific world. In their introduction the editors recount the story of my Damascene conversion to process ontology in Amsterdam. Though this is as true as such stories can be, conversion requires a good deal of preparation of the soil.

My preparation for embracing process ontology was undoubtedly my engagement, especially in collaboration with Maureen O'Malley, with microbiology and the phenomena therein of symbiosis and lateral gene transfer. Traditional substance ontology sees the world as composed of distinct things with more or less definite boundaries. Contemporary microbiology, and the world of symbiosis that it has revealed to us, sees systems with ever-changing parts. Countless microbes flow in and out of the "holobionts", multicellular organisms and their microbial associates, and the microbes themselves exchange plasmids and similar vehicles for genetic material. Every "thing" flows.

Gilbert applies an encyclopaedic knowledge of both the history of biology and contemporary biological science to explaining the nature of symbiosis and its relevance to our general understanding of life as a process. His chapter beautifully weaves together poetic images of the various interwoven flows of living matter over time scales from the biochemical to the evolutionary with fascinating details of the ways in which microbial and macrobial life interact. I would happily refer anyone who didn't understand the significance of process philosophy of biology (or life) to this chapter.

Doolittle and colleagues (Papale et al., Chapter 6 of this volume) address the profound implications of symbiosis for evolution. My friends and colleagues have often chastised me for speaking too much of ontology and too little of epistemology. I have recently begun to try to address this, notably in a paper with Leonelli (Dupré & Leonelli, 2022). But it is especially gratifying to see process philosophy contributing to real scientific debates. Papale et al. explicitly describe their chapter as process epistemology and remain agnostic about the underlying ontology. But for me, anyhow, the success of process epistemology provides further evidence for the necessity of a process ontology to explain this success and also, thereby, to motivate further efforts to develop process epistemologies.

This chapter makes a compelling case that Doolittle's much-discussed "it's the song not the singer" account of evolution, interpreted as a process theory, solves the taxing problem of how to make evolutionary sense of the pervasiveness of symbiosis. Philosophers of biology will also be very interested in the argument that this idea also favours a replicator/interactor view of evolution over the widely accepted view, associated with Richard Lewontin, of heritable variation in fitness. The lack of simply describable inheritance between holobionts, the sum of a macrobe and its associated microbial symbionts, has generally been seen as a fundamental objection to seeing holobionts as units of selection. But surely, they did evolve; we did, for example. So a theory of evolution that makes this possible is highly desirable; Doolittle and colleagues show that it is also entirely possible.

Pradeu comes to the importance of symbiosis and microbiology from a different direction, immunology, revisiting his influential arguments that the immune system cannot be adequately understood in the traditional way as distinguishing self from son-self. Central to this argument is the fact that the immune system does not treat countless symbiotic microbes as "non-self". He argues instead that the immune system responds to sudden changes in the chemical ambience of the organism, hence precisely addressing the question of stabilisation that is so central to a view of life as composed of processes. I concur happily with his judgement that our views have converged from very different directions. His account of immunology has provided a central plank of the rethinking of philosophy of biology towards cooperative interaction and flow and converges nicely with the account in Gilbert's chapter. Taken together, these three chapters provide a succinct summary of some of the most important scientific consequences of understanding biology as process.

Catherine Kendig's paper on soil extends this scientific impact to a more applied topic, and I was delighted to see a paper on soil in this collection. Soil is a place where several of my interests intersect, not least that of a keen gardener. It is also, as Kendig makes clear, a paradigmatic process. A few years ago I had a highly informative conversation with two soil researchers from the Rothamsted Institute, the main agricultural research centre in England. They reported to me that some of their colleagues were still inclined to treat soil as stuff that held plants upright and provided a route by which chemicals added to the soil could reach plants. This is probably a parody of what anyone really believed, but as Kendig makes clear soil is, on the contrary, an extraordinarily complex and dynamic process, involving a high degree of microbial interaction and cooperation. What I had not realised was what an excellent case study it provides for a promiscuous realist view of classification. Kendig nicely shows how different interests provide classifications in terms of physical characteristics such as particle size useful for engineering purposes and of a wide range of biological characteristics for environmental or agricultural purposes. The latter in turn lead to a variety of conceptions of plant health. The discussion illustrates the value of Kendig's concept of kinding, the activities of interaction with a category of entities that leads to locally useful taxonomies.

13.2 The Nature of Humans

I have always, and unashamedly, been concerned with the importance of science for human life, and several of the chapters in this book engage directly with this topic. Central to this concern has been the scientific impact on our understanding of what it is to be human, human nature. Tim Lewens takes up this topic. While his chapter is largely framed as an exposition, and ultimately largely an endorsement, of my own views on the topic, I learned a great deal from it. First, as I also noted about the introduction, I learned something about my own views, especially on how they had evolved over the past decades. I don't tend to read my own past work much, and if I do, not very carefully, so this careful exposition was fascinating, to me at least. But the chapter achieves much more than this. It not only adds arguments and telling examples, but also explains something about my views that I had not properly formulated. This is an exploration of the tension, which I certainly recognise, between scepticism about attempts to define human nature, and an acknowledgement that the concept can serve useful purposes, namely as a means of communication between different intellectual and wider communities. Lewens lucidly points out the tension, but also offers an account of why my tendency to equivocate here is, partially at least, justified.

Adrian Haddock, uniquely in this volume, comes from a background in mainstream metaphysics rather than philosophy of science. I must confess that I am delighted that he finds an important insight in Chap. 7 of *The Disorder of Things*. Though that book has received as much discussion as I could reasonably have hoped, I'm not aware of any other serious reflections on this chapter. I do remember, however, being quite pleased with it, and having finally reread it just before writing this, I still like it. Moreover, while I can't say exactly what I thought I was doing over 30 years ago, I find myself very much liking Haddock's account of it.

But of course, Haddock's kind remarks on that chapter are something of a Trojan horse. What he claims I have shown, in much too brief summary, is that a theory of the mind cannot be an empirical theory in the way that the scientific theories we believe are. And he goes on to claim that this insight contradicts the empiricism to which I am also committed and, as I read him, especially the naturalistic metaphysics to which I regularly claim allegiance. A proper response to this would require at least a chapter. It calls on me to reassess or at least further elaborate my most general views about the nature of metaphysics and epistemology and what I mean by "empiricism".

However, I can offer a partial response.[1] Haddock defines empiricism as "the idea that the meaningfulness of language consists in its containing empirical information". This is not what I mean by the term or what I believe. My empiricism is no more than the claim that knowledge of what the world is contingently like must be grounded in experience. It is, in this sense, an empirical fact that we are a species with a rich

[1] And, incidentally, my naturalistic metaphysics is not intended to advocate an end to the more traditional subject. Indeed, I recently co-edited a book, with Anne-Sophie Meincke (Meincke and Dupré 2020), intended to bring traditional metaphysics in constructive contact with more naturalistic philosophy of biology on the topic of biological individuality.

and complex language including, for instance, a rich and occasionally pretentious vocabulary for describing wine. Haddock and I agree that this fact presents a serious difficulty for a reductive account of the mental. It is surely a contingent, empirical fact that we have a language that has the resources Haddock discusses, but for him there are momentous metaphysical consequences of this fact that I fear I must, if reluctantly, disown.

The deepest difference between Haddock and myself, I suspect is in how we understand metaphysics. Regarding "my attempts to synthesise the results of empirical (and specifically, biological) science into a comprehensive theory of the world", he writes: "that is what it is: a theory … a body of empirical beliefs. It is not the sublime activity that characterises the philosophical tradition: the articulation of the fundamental form of reality." I do think that we can have empirically well-grounded accounts of "the fundamental form of reality", but I fear this is a less sublime conception of metaphysics than that to which Haddock alludes. In summary, the resolution of the disagreements between us would require a thorough examination what each of us means both by empiricism and by metaphysics.

There was as stage in my career in which I thought the most important task for philosophy of science was a kind of intellectual hygiene, particularly in relation to areas of science that purported to offer accounts of human life that I considered both simplistic and dangerous. My specific targets were evolutionary psychology and economics. Nancy Cartwright is generous, however, in starting her chapter on economics with a quote from me. She has worked far more deeply than I have on the nature and limits of economic methods. But there are strong, and non-accidental, parallels between her work and my own—though in the case of economics the causal connection runs more from her than, as a naïve reader might guess from her starting point, the other way round. She is a member of what has come now to be referred to as the "Stanford School" with whom I overlapped for a substantial period at Stanford, and her work has certainly been a major influence on mine. The view of the world that I described as promiscuous realism is very similar to what she has called the dappled world. Both of us see the kind of axiomatic order that once was the standard view of a law-governed world as deeply mistaken: order is always local and partial.

The fundamental problem for the disordered world that both Cartwright and I believe is the reality in which we live is in understanding the indisputable success that science has provided in our understanding of it. This is what she elegantly does here for the case of economics. While she is quite deflationary in her account of economics, the fact that economics does not map the fundamental structure of reality, as some of its practitioners may still believe, does not show that it is entirely useless. Economic models suggest what Nancy once called ceteris paribus laws. But while this term worked well for the physical sciences, where experimental techniques can sometimes come close to making everything else relevant genuinely equal, or better absent, the diversity and complexity of factors involved in such phenomena as the economic (and, I would add, much of biology) makes the whole idea of a law unhelpful. Much better, as Cartwright does here, to think of tendencies that are captured by economic models. They don't literally describe the world, but they may help us in understanding the causes that operate in a particular part of it. However, taking them literally has been taken to justify policies that have been distinctly harmful to humans.

13.3 Metaphysics of Science

Jim Griesemer's and Philip Kitcher's chapters offer perhaps the most detailed critical engagement with my work in the volume, and this is not the place to provide an adequate response to either. I agree with Jim's summary: we agree about so much, yet somehow perhaps disagree about so much.[2] Jim suggests that some of the differences trace to our respective beginnings, mine in philosophy his in genetics. This may well be right; and perhaps he is also right that my philosophical origins have predisposed me to dualisms, such as unity/disunity and process/substance. The suggestion stings, however; in principle I have always thought two an absurd number in philosophy. One or many I can appreciate. But why two rather than 17 or 111? Perhaps my process monism is partly due to this distaste, as I couldn't see any options beyond the two of process and substance. I'll postpone discussion of why I am a metaphysical monist to my comments on Philip Kitcher's chapter.

Jim may well be right that I have been too harsh in some of my criticism of population level models; I admit these may have a legitimate place in science. (This resonates with Cartwright's qualified defence of economics.) On the other hand, he perfectly summarises the basis of my criticism in terms of missing levels. I also applaud his summary of the way forward: "Progress toward a multi-perspective philosophy of biology might be made by reimagining population- and gene-level modelling projects as a multi-scale problem that links (if not integrates) all three scales of molecule (and not just nucleic acids), organism, and group". I would only add that, at least from a process perspective, such linking faces the central challenge of dealing with the fact that these exist at very different time scales.

I also accept the charge of being a representationalist, at least in so far as this just means that I am a realist: I think scientific beliefs are often true because they represent the way things are. On the other hand, I have much sympathy with the idea that models might be more like navigational tools, good for steering us through complex domains. I am happy to be an epistemological pluralist. Are these different perspectives that can be applied to the same models, or specific perspectives offered by different kinds of models? I don't know, but I'd like to. Jim again takes a different path here, following William Wimsatt into an onto-epistemological thicket. Do I want to go there? It sounds an easy place to get lost in. I'll hope to sketch some sort of response in my discussion of Philip's chapter.

[2] An unusual problem in writing the chapter has been how to refer to the contributors. In general other contributors have referred to me by my last name in standard academic style, though occasionally they have used my first name in more personal contexts. I decided generally to follow this convention. However, the last two contributors, Jim Griesemer and Philip Kitcher refer to me throughout by my first name, and I shall do the same. All the contributors are good friends whom I would not think of addressing by their last names in other contexts.

Although Philip's pragmatism is a bit more familiar than Jim's onto-epistemological thicket, they have much in common. Philip ends, after all, speaking of "worlds half made", suggesting worlds inextricably entangled with epistemology. Philip's strategy is to applaud my early promiscuous realism, and to deploy it against the process ontology of my later work. He has powerful arguments, but I remain unrepentant.

Why be a metaphysical monist, having spent half my career defending pluralism? Why do metaphysics at all? Philip wants to follow Pierce's wise advice that we ought to abandon questions that lack implications for human conduct. But I insist that metaphysics does matter for science and science matters for human conduct. Philip might well agree with this. But he will, no doubt, continue, why not whatever metaphysics helps with the particular problem at hand? Why not promiscuous, pragmatist metaphysics? Why not a metaphysics "half-made" for the purpose at hand?

I have, as Philip surmises, always been suspicious of the idea that ultimately in science we make things up. I did indeed embrace something like the view of reality Philip likens to a piece of marble full of seams. I don't feel moved by the objection that there is (almost) no limit to the number of seams we consider "natural". Surely the number of distinctions we make in nature is vastly less than the number that we could recognise in theory. But no doubt that number will increase as new human projects emerge, though others will equally come to seem less interesting.

But with the turn to process I reject this metaphor: marble is too stable, thing-like. I rather see the world as an ever-changing process in which temporary boundaries appear, are stabilised for a time, and disappear. Much of this stabilisation, especially in science, is done by us. We don't make things up; we make things. Much of the work in a laboratory is making things by stabilising living processes in particular ways; but always only temporarily. This has proven a good way, up to a point, of helping us understand how natural processes stabilise themselves. Philip thinks we want "to say what is true—or at least true enough to advance our purposes". I agree. But the limitations on this goal are not the Kantian inscrutability of a noumenal world, but the reality that the world is ever-changing and whatever we say about it is liable to be true only for a limited time. We develop our language not only to get closer to the truth, but to track an ever-changing reality.

This matters because we are always disposed to interpret the way we speak as describing a world of things and the way they are. Perhaps this tendency is so deep a part of our biological nature, or so deeply entrenched a part of our culture and psychology, that we must learn to live with it. But as Leonelli and I argued, this can lead us astray in serious ways (Dupré & Leonelli, 2022). Recognising the things in terms of which we see the world as reifications—temporarily stabilised things—can help us to be aware of the dangers of this erroneous metaphysics. There is much more to be said here. I hope, at least, that Philip and I will remain stable enough for a while to continue the conversation.

13.4 Interdisciplinarity

The final section of the book is concerned with interdisciplinarity. Interdisciplinarity has been strongly promoted by university administrators and funders for all of my academic career, but the structure of universities has generally discouraged it. This is a good point at which to express my gratitude to the UK Economic and Social Research Council (UK) that provided the considerable funding that made Egenis possible. It is sadly ironic that the day before writing this I saw a message from UK Research and Innovation announcing that they were cancelling the explicitly interdisciplinary scheme that was funding my main current interdisciplinary project, involving myself, an ecologist, an artist and a sociologist.

That very sociologist, Jane Calvert, proposes, in her chapter, that rather than studies of science, we should consider studying with science. Taking this thought one step further, she proposes studying with other organisms rather than just studying them. As for the first, I think philosophers are not quite as far behind as Calvert sometimes suggests. I think especially of Pradeu's prepositional innovation, his advocacy of the move from philosophy of science to philosophy in science. But the collaborations she describes are undoubtedly enlightening and I don't know that any philosophers have explicitly made the second move.

I was struck by a remark Calvert makes in relation to her collaborative work with synthetic biologists on re-engineering yeast. As she notes, with reference to the early synthetic biologists' attempts to treat the parts of organisms like Lego bricks: "But living systems tend to resist these attempts at standardisation." I was reminded of Alex Blanchette's brilliant but disturbing book, Porkopolis: American Animality, Standardized Life, and the Factory Farm (Blanchette, 2020). Blanchette describes the ambitions of the factory farm managers to standardise their pigs, which would, among other things, enable their automated "disassembly" in the final stage of their lives. Happily, like Calvert's yeast, the pigs have so far resisted this standardisation.

As both these examples show, treating living systems as mechanistic assemblages of parts does not align well with the nature of life. And, as Calvert agrees, a process ontology can explain this failure and even suggest better ways forward. I suppose that a traditional pig farmer might well present a good example of working with animals. While the pigs would probably not have embraced the final goal of ending up as sausages, surely up to this point the farmer could be seen very much as collaborating with the pigs in keeping them healthy, comfortable and (very) well-fed. Perhaps traditional animal husbandry might be a source of inspiration for the possibility of science with organisms?

The conference out of which the chapters of this book grew was not solely focused on my philosophical work. On the contrary, two days out of three were concerned with the achievements and future prospects for Egenis, celebrating its 20th anniversary. So it is very appropriate that Alan Love's chapter focuses specifically on the virtues of interdisciplinary centres and the parallels between Egenis and the centre he directs in Minnesota. Love describes an impressive large scale interdisciplinary project he has directed on teleology and the importance of intellectual community. This I can only admire; I'm not sure I have the kind of managerial skills required to organise a project of this complexity.

I am, however, very proud of whatever I have contributed to Egenis, and the second of Love's themes captures exactly what I am proud of. Egenis is a real intellectual community in just the way Love describes and, as he notes, these are not so common nowadays and generally are far removed from the aims of university administrations. This leads me to reflect on what ingredients are apt to make intellectual community happen. I suggest three: people, space and probably luck. If I have a skill as a manager, it is surely with the first. An intellectual community requires talented thinkers who are open-minded, interested in listening to different perspectives and projects, and interested in talking to one another on intellectual topics. My great success (and good fortune) has been in bringing together a substantial number of such people.

Space is also essential. I was fortunate to receive advice from a geographer, Michael Winter, at the outset of Egenis, who impressed upon me the necessity of a physical location for a Centre, its centre. Virtual communities have their role, no doubt, but they cannot come close, in my experience, to the kind of community that Love describes in Minnesota, or which I have enjoyed at Egenis. Winter's advice was at the forefront of my mind when I received a generous private donation to the Centre soon after its foundation. Several people advised me to use this to appoint postdocs or PhD students; but my decision was to acquire long-term use of our beautiful building, Byrne House. It is a decision I have never regretted.

Luck, I fear, is essential for all projects, and both the factors mentioned already require an element of that. I often remember a probably no more than half-serious remark by my former Co-Director, sociologist of science Barry Barnes, who told me that his strategy in life was to find a lucky person and attach himself to them. My career has had many strokes of luck, among which finding Barry and Steve Hughes, a genomics scientist, my other original Co-Director, to help launch Egenis was not the least. It is richly exemplified by the wider group of colleagues and dear friends among whom are the contributors to this volume.

Finally, on the question of management, Egenis has grown over the years to a size that has very probably outgrown my managerial skills. In the last few years it has been fortunate indeed (though I will again claim some credit in regard to my original personnel selection skills) to have found two successors, Sabina Leonelli and then Adam Toon, who have no such deficits. Under their leadership Egenis has acquired clearer structure and organisation, while maintaining and enhancing the kinds of practice that contribute to the inclusive community that Love describes. I believe it is in great shape to continue to grow and flourish.

* * *

Philip Kitcher begins his chapter with some of my professional and even personal history and some very generous remarks about these. I am English so, as he says, a predominant reaction is inevitably embarrassment—but embarrassment of a very pleasant kind. I thank him and all the other contributors who have said similarly generous things about me. But most of all I want to thank all the contributors for years of friendship and for their engagement with my work. I would like to think that this is a point at which my practice echoes my philosophical views. Much of my work has stressed cooperation over the too common obsession with competition, most notably in economics and evolutionary biology. Competition is an important phenomenon, but life is everywhere partly cooperative (Dupré & O'Malley, 2009). In my recent book *Everyone Flows* (Dupré, 2025) a central theme was the uniquely cooperative and independent nature of humans. All the human accomplishments and achievements that we admire depend for their existence on a host of other humans providing the conditions of life, the professional tools, and the historically acquired knowledge of countless others.

Philosophy is no different. Though it is sometimes thought of as the work of the isolated genius, seeing more deeply into things than others, this is a misleading perspective. It is not only that philosophers need to be raised, fed, clothed, housed and transported about like anyone else, though this is certainly true and important. All good philosophy also depends on the insights of countless others, many now dead. But finally, the practice of philosophy itself is at its best in the kind of intellectual community I have tried to describe above. Perhaps what pleases me most about this book is that it represents just this. All of the contributors are scholars from whom I have learned much, and some are kind enough to say that they have learned something from me. Several of them are, or have been, immediate colleagues in Egenis, but of course this book does not reflect just this one intellectual community, but the intersection of many, a transient community of communities. But that, I think, is philosophy at its best.

In summary, there can be nothing more gratifying after a life of grappling with intractable philosophical problems than to have many of the colleagues you admire most devote some of their precious time to reacting to your thoughts. So thank you!

References

Blanchette, A. (2020). *Porkopolis: American animality, standardized life, and the factory farm*. Duke University Press.

Dupré, J. (2025). *Everyone flows: A process philosophy of human life*. Oxford University Press.

Dupré, J., & Leonelli, S. (2022). Process epistemology in the COVID-19 era: Rethinking the research process to avoid dangerous forms of reification. *European Journal for Philosophy of Science, 12*(1), 20.

Dupré, J., & O'Malley, M. A. (2009). Varieties of living things: Life at the intersection of lineage and metabolism. *Philosophy, Theory, and Practice in Biology, 1*, 20130604. (2009).

Meincke, A. S., & Dupré, J. (2020). *Biological identity*. Routledge.

MIX
Papier aus verantwortungsvollen Quellen
Paper from responsible sources
FSC® C105338

If you have any concerns about our products,
you can contact us on
ProductSafety@springernature.com

In case Publisher is established outside the EU,
the EU authorized representative is:
Springer Nature Customer Service Center GmbH
Europaplatz 3, 69115 Heidelberg, Germany

Printed by Libri Plureos GmbH
in Hamburg, Germany